T0313370

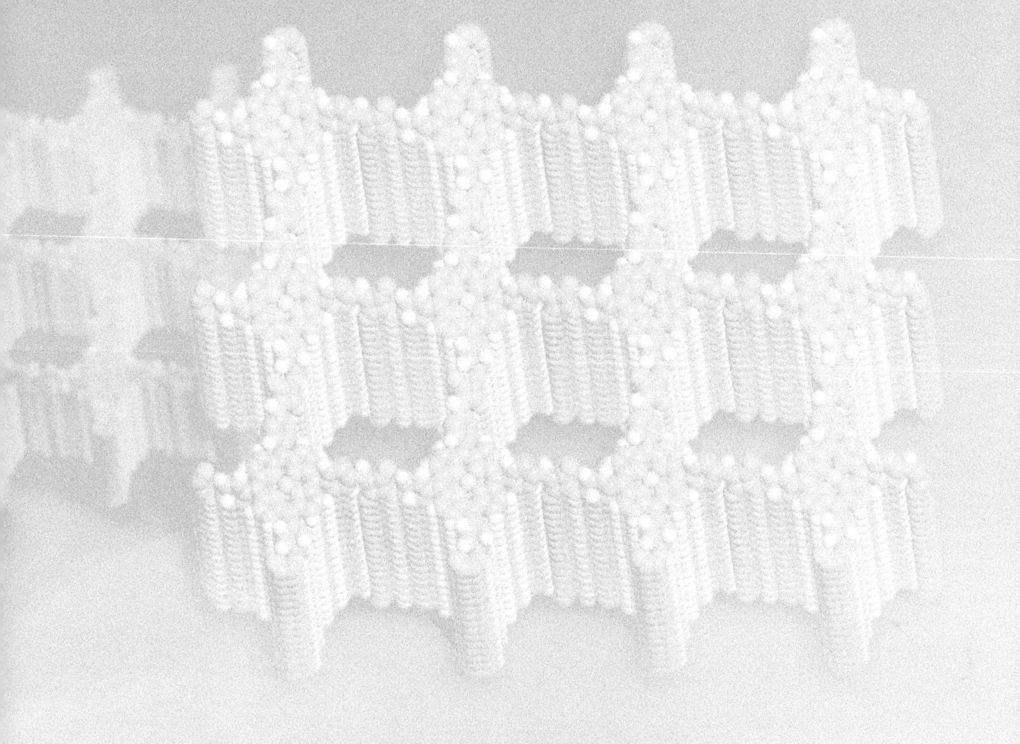

Covalent Organic Frameworks

Covalent Organic Frameworks

edited by

Atsushi Nagai

Published by

Jenny Stanford Publishing Pte. Ltd.
Level 34, Centennial Tower
3 Temasek Avenue
Singapore 039190

Email: editorial@jennystanford.com
Web: www.jennystanford.com

British Library Cataloguing-in-Publication Data
A catalogue record for this book is available from the British Library.

Covalent Organic Frameworks

ISBN 978-981-4800-87-7 (Hardcover)
ISBN 978-1-003-00469-1 (eBook)

Contents

Preface

Rational synthesis of extended arrays of organic matter in bulk, in solution, in crystals, and in thin films has always been a paramount goal of chemistry. The classical synthetic tools to obtain long-range regularity are, however, limited to noncovalent interactions, and not covalent polymerization reactions, which usually yield structurally more random products. The most challenging hurdle in the synthesis of extended yet precisely defined 2D and 3D structures based on covalent chemistry is the requirement that the reaction linking individual organic constitutes be reversible, allowing the scaffold to arrange into a thermodynamic, well-ordered product rather than a kinetic, amorphous structure. Although the synthesis of crystalline inorganic materials largely relies on dynamic polymerization processes, only a few examples that use dynamic polymerizations for the generation of crystalline organic frameworks are known.

Hence, a combination of porosity and regularity in organic covalently bonded materials requires not only the design of molecular building blocks that allow for growth into a nonperturbed, regular geometry, that is, two or three latent bonding–forming sites spanning an appropriate angle, but also a condensation mechanism that progresses under reversible, thermodynamic, self-optimizing conditions. Ideally, a single precursor molecule fulfills all these criteria, making even (if any) known from step-growth polymerization reactions such as polycondensation and polyaddition. In 2005, crystalline covalent organic frameworks (COFs) as 2D polymers were for the first time reported by Yaghi et al. 2D COFs resemble an sp^2-carbon-based graphene sheet, but their structures have a different molecular skeleton formed by orderly linkage of building blocks to constitute a flat organic sheet. Nanocarbons are also intriguing molecules and motifs with a discrete size and 2D conformation but beyond the scope of this book. In the past decade, COFs have emerged as a new class of highly ordered crystalline organic porous polymers. They have attracted tremendous research interest because of their unique structures and potentially wide applications in gas storage and separation, energy storage, catalysis, and optoelectronic materials development.

This book, *Covalent Organic Frameworks*, with a selection of subjects, is a handbook of COF research from design to application, providing basics of and current trends in organic porous materials for successful development and application and describing the concepts of COF design and synthesis; COF crystallization and structural linkages; theory of gas sorption; and various applications of COFs, such as heterogeneous catalysts, energy storage (such as semiconductors and batteries), and biomedical approaches.

Atsushi Nagai
Delft University of Technology (TU Delft), Netherlands
2020

Chapter 1

Design and Synthesis: Covalent Organic Frameworks

Covalent organic frameworks (COFs) are newly emerged crystalline porous polymers with well-defined skeletons and nanopores mainly consisting of lightweight elements (H, B, C, N, and O) linked by dynamic covalent bonds. Compared with conventional materials, COFs possess some unique and attractive features, such as a large surface area, a predesignable pore geometry, excellent crystallinity, inherent adaptability, and high flexibility. Their large surface area, tunable porosity, and π conjugation, with unique photoelectronic properties, will enable COFs to serve as a promising platform for a wide variety of applications. This chapter traces the evolution of COFs and highlights the important issues related to synthetic method and structural design.

1.1 Introduction

Recent decades have seen an explosion of interest in organic materials with permanent nanometer-scale pores that grow very quickly due to their specific properties and have broad applications in gas storage, gas separation, drug delivery, energy conversion, catalysis, and optoelectronics. Until now, a large number of typical porous materials have been designed and constructed, such as

---

Covalent Organic Frameworks
Atsushi Nagai

ISBN 978-981-4800-87-7 (Hardcover), 978-1-003-00469-1 (eBook)
www.jennystanford.com

zeolites, metal-organic frameworks (MOFs), mesoporous silica, organosilica, microporous polymers, porous carbons, covalent organic frameworks (COFs), and organic molecular cages. Among them, the covalent chemistry of organic and inorganic molecules has been the focus of chemists, leading to many important advances in science. The building and modification of organic molecules by covalent bonds to make pharmaceuticals, chemicals, and polymers have fundamentally changed our way of life. Likewise, covalent synthesis through the construction of inorganic complexes has led to useful catalysts capable of high activity and selectivity. The precision and versatility with which covalent chemistry on such molecules is practiced have not been translated to either the building units of extended structures or their modification. This is the next-generating idea of MOFs [1] and COFs [2–4], which are two popular classes of porous materials on account of their unique properties and promising application potential, attracting a lot of research attention. Constructing novel MOFs and COFs is necessary to expand the family of porous materials. This book focuses on COFs. Generally, a 2D COF that is a covalent 2D polymer would resemble a sp^2-carbon-based graphene sheet but its structure would have a different molecular skeleton, formed by orderly linkage of building blocks to constitute a fat organic sheet. Two-dimensional COFs are not single-atom-thick sheets but consist of 2D sheets layered as a π-π stack. Because the pore size and shape are determined by the structure of the building blocks, 2D COFs represent a class of porous materials with tailor-made components and compositions. This chapter is divided into three sections, which focus on the design and synthesis of COFs.

1.2 Design and Synthesis

1.2.1 COF Synthesis in the Dynamic Covalent Chemistry Concept

The crystallization of COFs has generally been dominated by kinetically controlled reactions that form irreversible covalent bonds. Higher reversibility, that is, dynamic covalent chemistry (DCC), must be achieved for efficient error correction [5]. DCC is thermodynamically controlled and thus offers reversible reaction systems with "error checking" and "proof-reading" characteristics, leading to the formation of the most thermodynamically stable structures. Figure 1.1a presents one such reversible assembly of molecular components, proceeding along kinetic or thermodynamics pathways. Due to the lower barrier of activation ($\Delta G^{\ddagger}$), kinetic intermediates dominate initially due to a fast rate of formation of the kinetic products, which can be trapped during quenching of the reaction and form amorphous products with a gain of entropy from their randomness of arrangement. Meanwhile the thermodynamic products have the lowest overall Gibb's free energy (ΔG°), which gives the opportunity for the reaction to re-equilibrate toward the global minimum to form the most stable and probably crystalline products. In contrast, kinetic products can also be transformed into thermodynamic products by using certain thermotreatments for recrystallization. Thermodynamic products have in common COF chemistry of the amorphous products initially obtained until the synthetic conditions are optimized by varying the crystallization time, reaction temperatures, reaction solvents, and catalysts (such as acids, bases, and metal salts acting as Lewis acids). Simply put, on using the DCC concept for the construction of COFs, a polymer skeleton is formed alongside the crystallization process, while the self-healing feedback reduces the incidence of structural defects and assists in the formation of an ordered structure. As a result, the final COF product possesses an ordered crystalline structure with high thermodynamic stability. Figure 1.1b shows the assembled polymer formation states of products between transition state and crystalline products; first the building blocks prepare hyperblanching polymers or cross-linking polymers. Then pore formation occurs with some pore errors. Finally, error correction using the DCC concept helps form a layer with no pore errors and the COFs are precipitated as crystalline products in the reactive solution.

On the other hand, rational design of building blocks may also increase the correctness of the assembly of monomer components, enhancing the crystallinity of the obtained products. It is generally believed that 2D COFs are easier to form, having less probability of resulting in networks than the 3D ones, with more structural diversity.

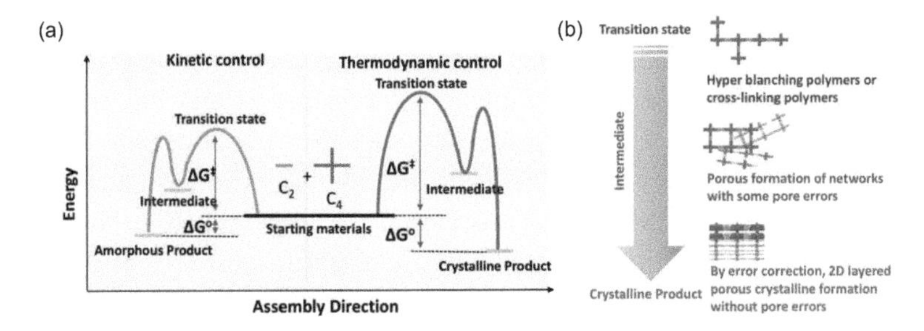

Figure 1.1 (a) Thermodynamic control in the dynamic covalent chemistry of COFs and (b) assembly state of COFs between thermodynamic control.

1.2.2 Dynamic Linkages of Building Blocks

One of the most significant characters of COFs is their periodic 2D or 3D polymeric networks linked by dynamic covalent bonds (DCBs), which are generally formed through reversible organic reactions as described above. Different from traditional interactions, DCBs possess robustness of covalent bonds and error-checking capability of reversible bonds at the same time [6], which endow COFs with great thermostability, multifarious adaptiveness, as well as considerable crystallinity. COFs exhibit various performance with different dynamic linkages. For example, robust linkage will enhance their chemical stability at the expense of crystallinity since there is a tradeoff between the two properties [7]. Table 1.1 presents a list of typical dynamic linkages and their structures, stability, and crystallinity.

In 2005, the Yaghi research group synthesized two microcrystalline 2D frameworks [8], COF-1 and COF-5, by the self-condensation of 1,4-benzenediboronic acid (BDBA) and co-condensation of BDBA with hexahydroxyltriphenylene (HHTP). Further, borosilicate [9], borazine [25], and ionic spiroborate linkages [10] were used for the formation of COFs. Lewis acid/base interactions of boronate esters with nitrogen and oxygen nucleophiles [27] were utilized for the formation of supramolecular frameworks [28].

For imine condensation, the Yaghi research group pioneered both imine [11, 29] and hydrazine [12] functionalities for the construction of COFs. Furthermore, azine [17] and imide [18] linkages have been utilized for the synthesis of COFs. The cyclotrimerization of cyano groups can lead to covalent triazine frameworks (CTFs) under ionothermal conditions [21]. In most cases, however, amorphous materials are obtained, especially at higher temperatures [30].

Several strategies have been developed to increase the hydrolytic stability of self-assembled materials based on C=N bonds. The Banerjee research group used 1,3,5-triformylphloroglucinol (Tp) as a trigonal building block for COFs [15]. This resulted in the irreversible conversion of reversibly formed enol imines into the more stable β-ketoenamines. Framework TpPa-2 [15] (Pa-2: 2,5-dimethyl-*p*-phenylendiamine) showed exceptional stability toward both the acidic 9 N HCl and the basic 9 N NaOH. In another study, however, Dichtel and coworkers showed that such linkages can still undergo dynamic exchange with amines and attributed the stabilizing effect merely to hydrogen bonding instead of irreversible tautomerization [31]. The stabilization of imine by intermolecular hydrogen bonding was shown for the porphyrin-containing COF 2,5-dihydroxyterephthalaldehyde–5,10,15,20-tetrakis(4-aminophenyl)-21*H*,23*H*-porphyrin (DhaTph), whereby Dha was utilized as a linear linker [32]. Samples of DhaTph remained stable even after suspension in water for more than 7 days. Quantitative conversion of imine linkages into the more stable amides resulted in the improved stability of the COFs [14], which was stabilized by the irreversible reduction of imine to the more stable amide.

Other types of dynamic covalent reactions have been utilized to a much lesser extent. Conjugated COFs were obtained by Michael addition elimination reactions [16] or the condensation of aldehydes with benzylic nitriles [24], with the latter approach resulting in an all-carbon framework. The polycondensation of squaric acid with a tetraaminoporphyrin led to the formation of a crystalline squaraine framework [19]. Wuest and coworkers utilized the reversible dimerization of the nitroso group to form azodioxides in the synthesis of single-crystalline 3D COFs [26].

This section summed up briefly what kinds of dynamic covalent linkages are utilized for the preparation of COFs. In Chapter 2, the COFs that build typical bonding structure are presented in detail by distinguishing DCBs as B–O (boroxine, dioxaborate, and spiroborate bonds) and C–N (imine, hydroazone, azine, squaraine, imide, phenazine, imidazole, and triazine bonds) linkages.

Table 1.1 Typical dynamic linkages for the construction of COFs

Bonds	Linkages	Characters	Ref.
B–O	Boronate ester 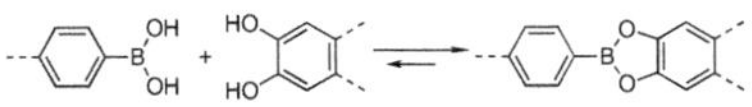	Formation temp.: 120°C Crystallinity: Excellent Thermostability: 600°C Chemical stability: Sensitive to water, acid, base, alcohols, and moisture	[8]
	Boroxine 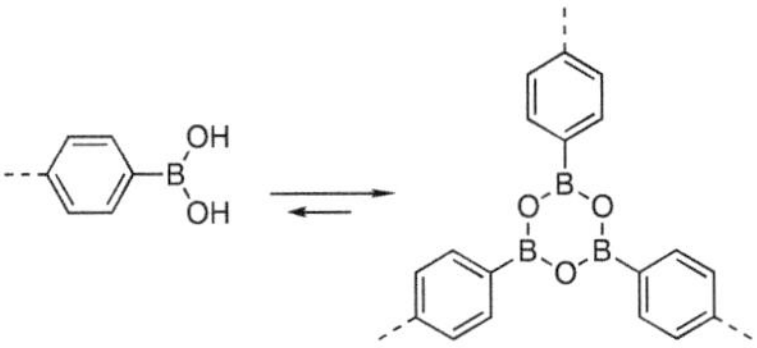	Formation temp.: 120°C Crystallinity: Excellent Thermostability: 500°C Chemical stability: Sensitive to water, acid, base, alcohols, and moisture	[8]
	Borosilicate 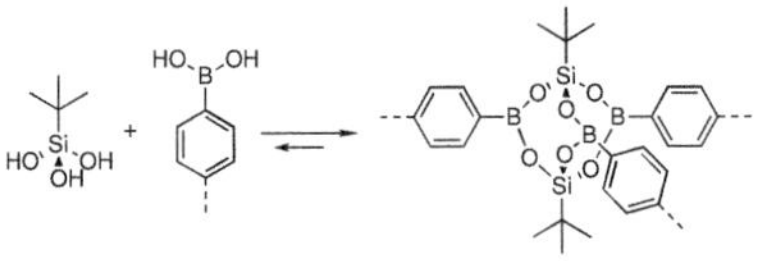	Formation temp.: 120°C Crystallinity: Excellent Thermostability: 450°C Chemical stability: For 1 h in air	[9]
	Spiroborate 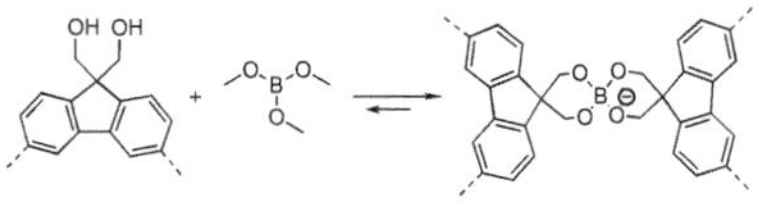	Formation temp.: 120°C Crystallinity: Excellent Thermostability: 400°C (weight loss 7%–12%) Chemical stability: 2 days in water and 1 M LiOH; sensitive to acid	[10]
C–N	Imine 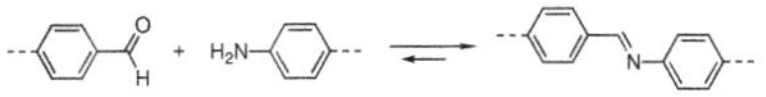	Formation temp.: 120°C Crystallinity: Good Thermostability: 500°C Chemical stability: Better than boron-based COFs	[11]
	Hydrazone 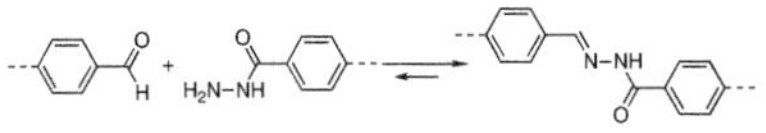	Formation temp.: 120°C Crystallinity: G Thermostability: 280°C Chemical stability: Better than imine-linked COFs	[12]

Bonds	Linkages	Characters	Ref.
	Imide 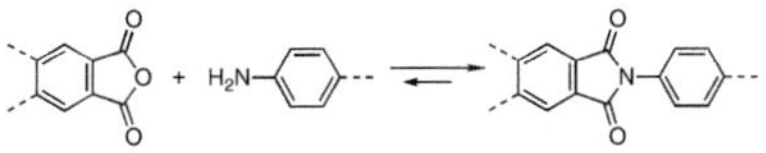	Formation temp.: 200°C Crystallinity: Good Thermostability: 530°C Chemical stability: Excellent	[13]
	Amide 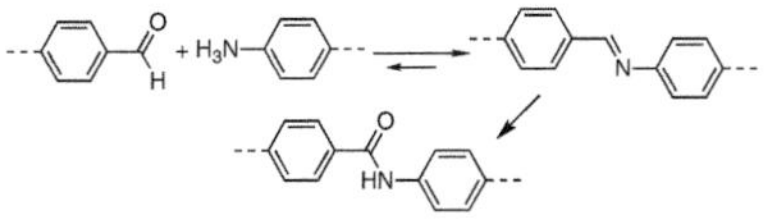	Formation temp.: 120°C/r.t. Crystallinity: Good Thermostability: 400°C Chemical stability: 1 day in 12 HCl and 1 M NaOH	[14]
	β-ketoenamine 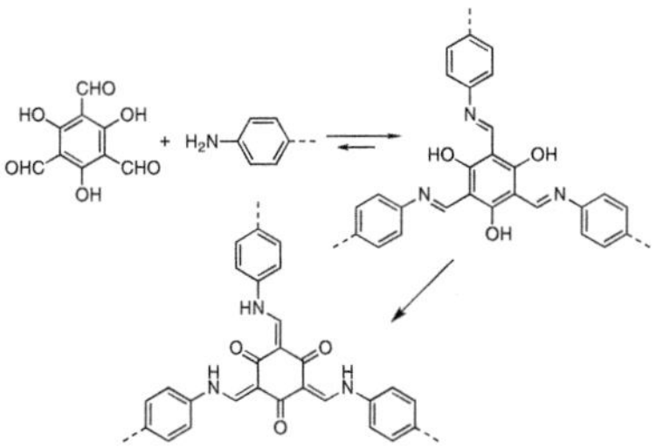	Formation temp.: 120°C Crystallinity: Good Thermostability: 350°C Chemical stability: More than 7 days in boiling water, 9 M HCl, and 9 M NaOH	[15]
C–N	β-ketoenamine 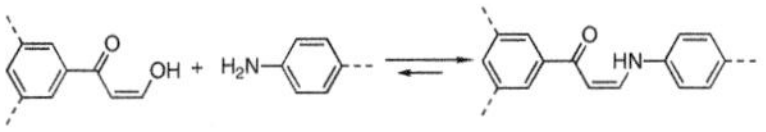	Formation temp.: 130°C Crystallinity: Moderate Thermostability: 300°C Chemical stability: 7 days in hot water (50°C) and 9 M HCl; partial hydrolysis in 9 M NaOH	[16]
	Azine 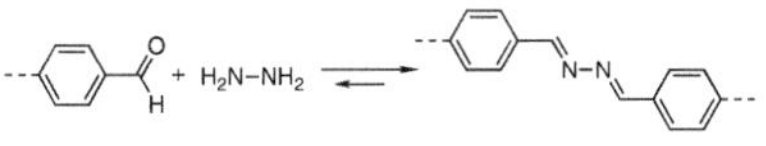	Formation temp.: 120°C Crystallinity: Moderate Thermostability: 300°C Chemical stability: 1 day in water, 1 M HCl, 1 M NaOH, and common organic solvents	[17]
	Phenazine 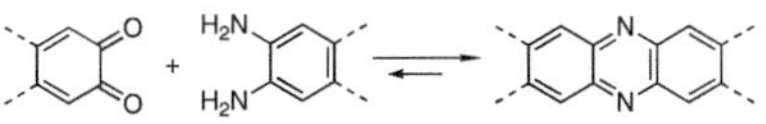	Formation temp.: 120°C Crystallinity: Moderate Thermostability: More than 1000°C Chemical stability: 1 day in water, 1 M HCl, 1 M NaOH, and common organic solvents	[18]

(Continued)

Table 1.1 (*Continued*)

Bonds	Linkages	Characters	Ref.
	Squaraine 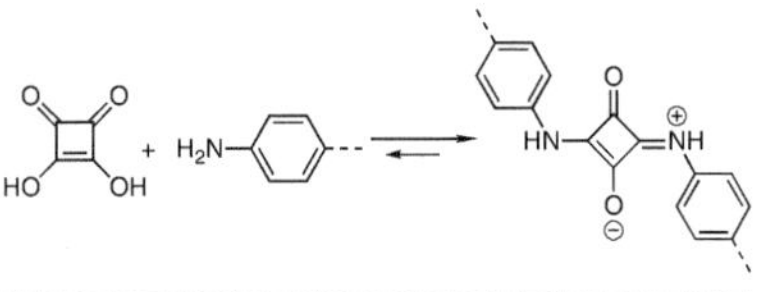	Formation temp.: 85°C Crystallinity: Moderate Thermostability: 300°C Chemical stability: 1 day in water, 1 M HCl, 1 M NaOH, and common organic solvents	[19]
	Viologen 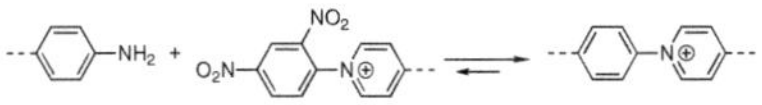	Formation temp.: 120°C Crystallinity: Moderate Thermostability: 400°C Chemical stability: 3 days in boiling water and 6 M HCl and sensitive to 1 M NaOH	[20]
C–N	Triazine 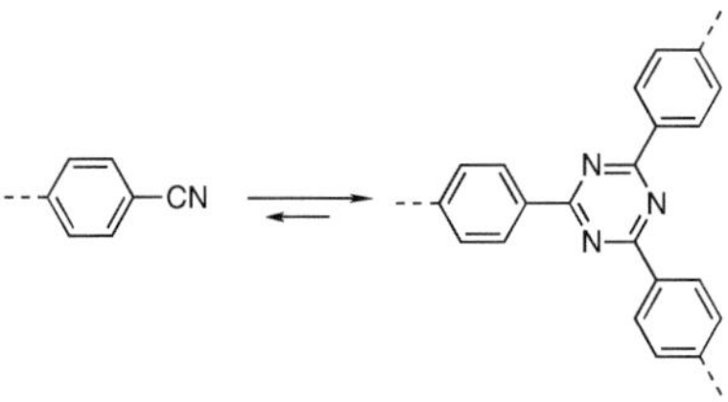	Formation temp.: 150°C Crystallinity: Poor Thermostability: 350°C Chemical stability: High (no precise description)	[21]
	Melamine 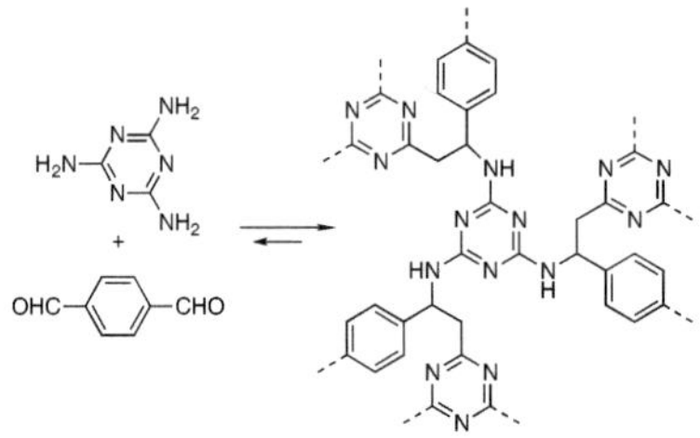	Formation temp.: Microwave Crystallinity: Poor Thermostability: 400°C Chemical stability: Stable in water and common solvents	[22]
C–C	C–C 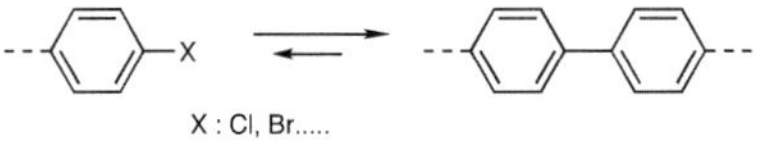 X : Cl, Br.....	Formation temp.: 300°C (annealing) Crystallinity: Not reported Thermostability: 250°C Chemical stability: Not reported	[23]
	C=C 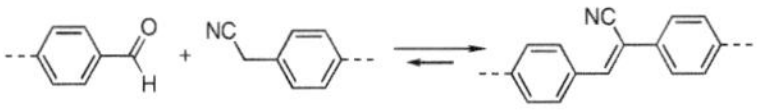	Formation temp.: 120°C Crystallinity: Moderate Thermostability: 250°C Chemical stability: High (no precise description)	[24]

Bonds	Linkages	Characters	Ref.
B–N	Borazine 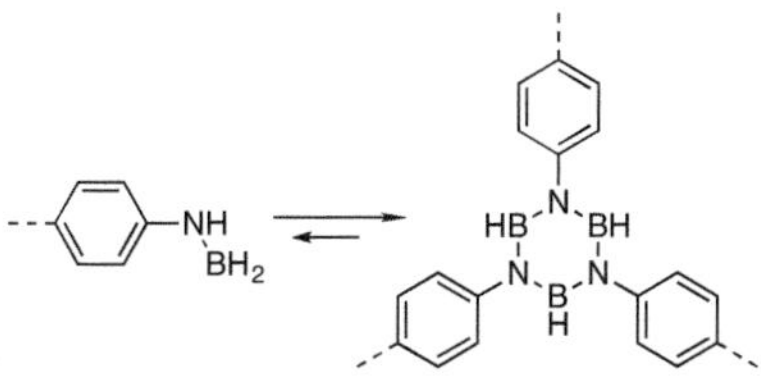	Formation temp.: 120°C Crystallinity: Good Thermostability: 420°C Chemical stability: No reported	[25]
N–N	Azodioxide 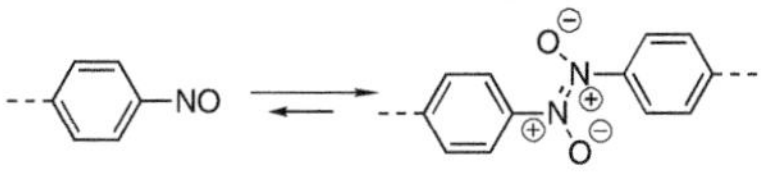	Formation temp.: 37°C Crystallinity: Monocrystalline Thermostability: Less than 130°C Chemical stability: Not reported	[26]

1.2.3 Topology and Geometry of 2D Porous Materials Containing COFs

Porous materials represent an important subject from a scientific point of view to study phenomena related to surfaces and confined spaces and have significant practical technological applications [33–35]. The design of methodologies to precisely control pore size and distribution to achieve a large surface area is a central theme in the exploration of porous materials containing COFs. For example, inorganic porous frameworks, the so-called zeolites, have recently attracted significant attention for their controllable porosity and systematically tailored chemical environment within the pores ranging from the micro- (<2 nm) to the mesoscale (>2 nm). Microporous zeolites, which are crystalline silicates or aluminosilicates, are typically synthesized under hydrothermal conditions from reactive gels in alkaline or acidic media at elevated temperatures [36–39].

In contrast to zeolites, MOFs with organic components in the frameworks can be synthesized using the coordination chemistry of metal ion bonds [40–44]. Thus, the formation of MOFs is driven by noncovalent coordination bonds [45–47], the diversity of metal ions, and the availability of different organic structures and topologies. The pore size can be tuned systematically from micro- or mesometer scales by using organic ligands of different lengths, while the topology of the framework is retained [48]. MOFs with large surface areas are very useful for gas storage and separation [49, 50]. MOFs have also been explored for their ability to act as catalysts for organic transformation [51] and polymerization [52] to show chirality [53], conductivity [54], luminescence [55], and magnetism [56] and to trigger spin transition [57] and nonlinear optical phenomena [58].

MOFs are unique in that they allow for topological design of porous structures because of the discrete angles and distances required for orbital overlap between metal ions and organic ligands. The discovery of MOFs has thus opened up new possibilities for designable molecular porous materials. However, one drawback of MOFs is that their structures are based on coordination bonds, which are much weaker than covalent bonds and make the structure fragile.

From the molecular design point of view, one of the significant characters of 2D COFs is the topological construction of the framework. This is possible because the geometries of building blocks determine the porosity and pore size. The connections between the precursors are built from covalent bonds and can be considered as rather rigid and an approximately linear connection between two subunits. Therefore, in this section, the topology and geometry of the assemblies are predominantly defined by both the number and the spatial distribution of reactive units in the molecular precursors. As a consequence of the rich chemistry of functionalized (poly)aromatic hydrocarbons, a great variety of planar precursors are available and numerous examples with C_2, C_3, C_4, and C_6 symmetries have been used to design and synthesize 2D COFs with various symmetries (Fig. 1.2). For example, condensation of C_3 monomers with C_2 monomers leads to 2D COFs with a hexagonal pore structure. On the other hand, condensation of the C_4 and C_2 monomers forms 2D COFs with tetragonal pores. Furthermore, triangular pore structures are formed by the condensation of C_2 with C_6 monomers.

To obtain 3D COFs, at least one building has to be extend in all three spatial directions. As a result of the maximum valency of four for sp^3-hybridized C atoms, simple 3D organic building blocks are to date exclusively based on tetrahedral tetraphenylmethane derivatives. The first work on 3D COFs was presented by Yaghi et al., in 2007, when they reported the successful formation of the 3D organic frameworks COF-102, COF-103, COF-105, and COF-108 for the first time by self-condensation of tetrakis-(4-dihydroxyborylphenyl) methane and its silane analogue tetrakis[4-dihyroxyborly]phenyl] silane and co-condensation with HHTP [4]. Overall, 3D COFs are still very limited in number compared to 2D COFs, presumably because of the increased difficulties and challenges in the synthesis of appropriate building blocks, crystallization, and structure verification.

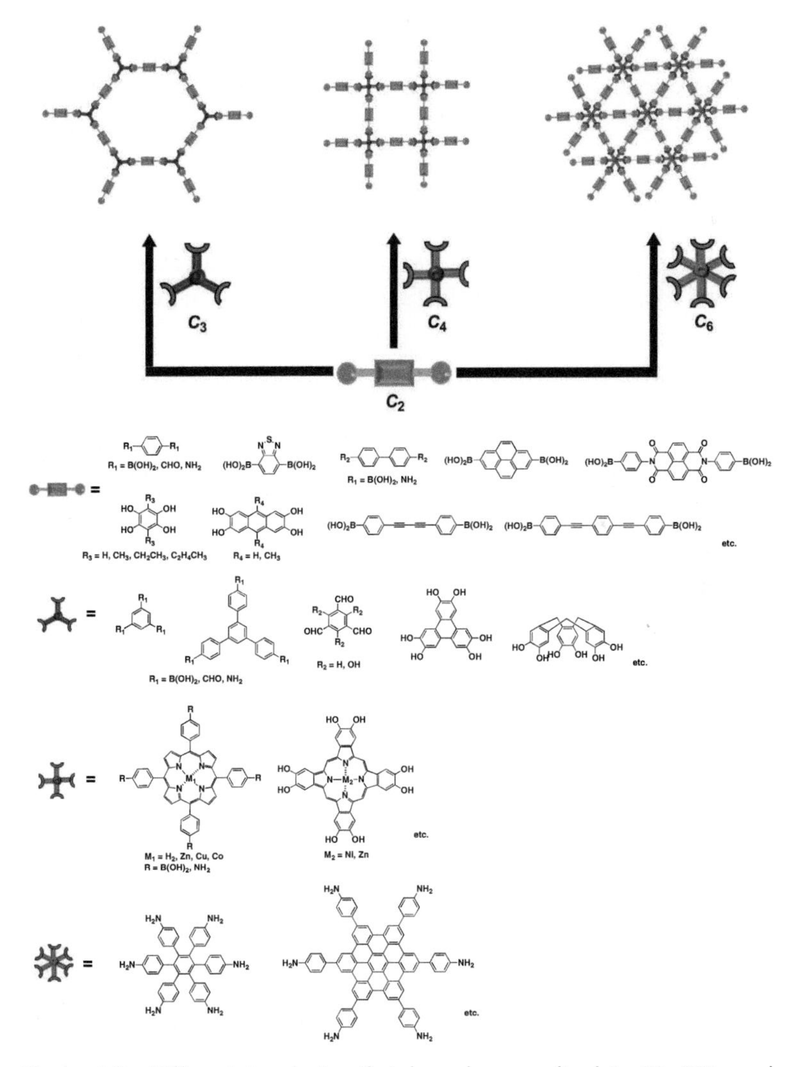

Figure 1.2 Different topologies that have been realized in 2D COFs and selected examples of planar building blocks for the formation of 2D COFs.

1.3 Synthetic Methods of COFs

Irrespective of the kind of material, a facile and productive method can be intriguing and makes it possible for the industrialization of production. Besides that, some special processes are also needed to build ideal structures. Given that, the building units and synthetic routes can be carefully selected on the basis of the design principles discussed above. Finding suitable synthetic conditions for COF synthesis is by no means a trivial issue. Since Yaghi and coworkers exploited the solvothermal (ST) conditions [3], many synthetic methods have been tried in order to search for a suitable way to satisfy the needs of extensive applications. Herein, the primary synthetic methods, such as ST synthesis [4, 11], ionothermal synthesis [21], microwave synthesis [59, 60], mechanochemical (MC) synthesis [61, 62], and room-temperature (RT) synthesis [63, 64], are discussed.

1.3.1 Solvothermal Synthesis

Most COFs are prepared via the ST synthesis method. A typical ST condition for the synthesis of COFs is as follows. Monomers and mixed solvents are placed in a Pyrex tube and degassed via several freeze-pump-thaw cycles. The tube is then sealed and heated to a designated temperature for a certain reaction time (takes 2 to 9 days). The precipitate is collected, washed with suitable solvents, and dried under vacuum to yield the COF as a solid powder. Issues and challenges such as solubility, the reaction rate, crystal nucleation, the crystal growth rate, and a self-healing structure are the most important points to consider when selecting the reaction media and conditions for the synthesis of COFs.

- Solvent combinations and ratios are important factors in balancing between framework formation and crystallization when synthesizing highly crystalline COFs.
- A suitable temperature is also important to ensure the reversibility of the utilized linkage reaction. Generally, COFs have been prepared at temperatures ranging between 80°C and 120°C, depending on the chemical reactivities of the building blocks. A closed reaction environment is required to allow the presence of water molecules, which could trigger the reverse reaction in the system.

1.3.2 Ionothermal Synthesis

Ionothermal synthesis was utilized to prepare triazine-linked COFs and was first reported by Thomas and coworkers [21]. Normally, aromatic nitrile as building blocks (e.g., 1,4-dicyanobenzene; 2,6-dicyanobenzene; and 1,3,5-tri(4-cyanophenyl)benzene) are dissolved in molten $ZnCl_2$ at 400°C and reacted for 40 h. In this process, $ZnCl_2$ acts as solvent as well as catalyst so the cyclotrimerization reaction seems to be partially reversible. When the system is cooled down to ambient temperature, the salts are washed out and crystalline CTFs are prepared after purification. However, harsh reaction conditions limit the building block availability. Most synthesized CTFs are amorphous materials that lack long-range molecular order.

1.3.3 Microwave Synthesis

Microwave synthesis has a long history in organic chemistry [65], but it was not until 2009 that Cooper and coworkers used this method to prepare COFs [59, 66]. The primal process was similar to ST synthesis. Microwave heating could be used to obtain 2D COF-5 and 3D COF-102 in 20 min., which is more than 200 times faster than the reaction time of 72 h required in the ST synthesis [8]. Moreover, the Brunauer–Emmett–Teller (BET) surface area of COF-5 (2019 m^2/g) obtained via microwave synthesis is slightly higher than that solvothermally synthesized in a sealed vessel (1590 m^2/g). Compared with ST methods, microwave heating allows the synthetic process to be completed in a faster and cleaner manner, providing a new possibility for further applications on larger scales.

1.3.4 Mechanochemical Synthesis

General preparations of COFs are used by dissolving building blocks in solvents, in which they can be limited by the physicochemical properties of monomers and the dissolving capacity of the solvents. The residual organic solvents may also feature toxic products [67]. MC synthesis, a traditional process technology on an industrial scale, may be an alternative to avoid these drawbacks [68, 69]. In the last few years, the MC method has been utilized in the preparation of porous organic polymers, including COFs [70].

In 2013, Banerjee and coworkers were first reported to have built β-ketoenamine-linked COFs by manual grinding in a mortar and pestle [61]. Interestingly, visual color changes could be observed and indicated the extent of polycondensation. After 45 min., the color of mixtures turned to dark red, which stood for complete COF formation (referred to as MC COFs). In comparison to the same COFs obtained from ST synthesis (referred to as ST COFs), MC COFs showed similar chemical stability but their crystallinity was moderate. To reduce such shortcomings, the liquid-assisted grinding method was applied to prepare imine, β-ketoenamine, and hydrogen-bonded imine-linked COFs [71]. Mechanical grinding can also be used to produce covalent organic nanosheets (CONs) from bulk COFs [72]. The prepared CONs retained their structural integrity and were stable in water, acids, and bases before exfoliation. Recently, Banerjee and coworkers developed a new MC method to continuously obtain COFs with a twin-screw extruder [62]. These COFs feature highly crystalline, ultrahigh porosity, and a high surface area. What's more, the COFs can be molded into any desired shapes and sculptures through heating, mimicking the ancient terracotta process. This approach is widely suitable for imine-linked COFs.

As mentioned above, CTFs were prepared by the ionothermal method in the beginning [21]. Later, the strategies of microwave synthesis and RT synthesis were developed with the help of a strong Brønsted acid catalyst by Cooper and coworkers. But the operating conditions and product yield were still unsatisfactory. Banerjee and coworkers expanded MC synthesis to construct CTFs [73]. Instead of the conventional cyclotrimerization of nitriles, the new approach is based on Friedel–Crafts alkylation.

In brief, MC synthesis is a user-friendly, ecofriendly, and time-saving alternative to produce COFs on a massive scale. It provides us a powerful method to realize the industrial production of COFs.

1.3.5 Room-Temperature Synthesis

RT synthesis was first mentioned in the review of Wang and coworkers, who first found that imine-based COFs could be facilely synthesized at RT in an ambient atmosphere [74], though their research paper was not published until 2017 [63]. By avoiding both the use of sealed vessels and the difficulty in controlling different synthetic parameters, this method makes bulk production of COF materials possible. The generality of this approach is under investigation. Further Zamora and coworkers used m-cresol and DNSO as a solvent instead of the traditional mixture solvent and prepared RT-COF-1 in minutes [75, 76]. Later, they combined this methodology with microfluid technology and constructed COFs in fibrillar microstructures [77]. When $Sc(TfO)_3$ was used to replace acetic acid, the COFs were synthesized within 10 min. and possessed a large specific surface area [78].

References

1. (a) Eddaoudi, M., Moler, D. B., Li, H., Chen, B., Reineke, T. M., O'Keefe, M., Yaghi, O. M. (2001). *Acc. Chem. Res.*, **34**, 319–330; (b) Kitagawa, S., Kitamura, R., Noro, S. (2004). *Angew. Chem. Int. Ed.*, **43**, 2334–2375.
2. (a) Ockwig, N. W., O'Keefe, M., Matzger, A. J., Yaghi, O. M. (2005). *Science*, **310**, 1166–1170; (b) Han, S. S., Frukawa, H., Yaghi, O. M. Goddard, W. A. (2008). *J. Am. Chem. Soc.*, **130**, 11580–11580.
3. (a) Hunt, J. R., Doonan, C. J., LeVangie, J. D., Côté, A. P., Yaghi, O. M. (2008). *J. Am. Chem. Soc.*, **130**, 11872–11873; (b) Côté, A. P., El-Kaderi, H. M., Furukawa, H., Hunt, J. R., Yaghi, O. M. (2007). *J. Am. Chem. Soc.*, **129**, 12914–12915.
4. El-Kaderi, H. M., Hunt, J. R., Mendoza-Cortez, A. P., Côté, A. P., Taylor, R. M., O'Keefe, M., Yaghi, O. M. (2007). *Science*, **316**, 268–272.
5. Rowan, S. J., Cantrill, S. J., Cousins, G. R. L., Sanders, J. K. M., Stoddart, J. F. (2002). *Angew Chem. Int. Ed.*, **41**, 898–952.
6. Jin, Y., Yu, C., Denman, R. J., Zhang, W. (2013). *Chem. Soc. Rev.*, **42**, 6634–6654.
7. Zhu, L., Zhang, Y. B. (2017). *Molecules*, **22**, 1149.
8. Côté, A. P., Benin, A. I., Ockwig, N. W., O'Keefe, M., Matzager, A. J., Yaghi, O. M. (2005). *Science*, **310**, 1166–1170.
9. Hunt, J. R., Doonan, C. J., Levangie, J. D., Côté, A. P., Yaghi, O. M. (2008). *J. Am. Chem. Soc.*, **130**, 11872–11873.
10. Du, Y. Yang, H., Whiteley, J. M., Wan, S., Jin, Y., Lee, S. H., Zhang, W. (2016). *Angew. Chem. Int. Ed.*, **55**, 1737–1741.
11. Uribe-Romo, F. J., Hunt, J. R., Furukawa, H., Klock, C., O'Keefe, M., Yaghi, O. M. (2009). *J. Am. Chem. Soc.*, **131**, 4570–4571.
12. Uribe-Romo, F. J., Doonan, C. J., Furukawa, H., Oisaki, K., Yaghi, O. M. (2011). *J. Am. Chem. Soc.*, **133**, 11478–11481.

13. Fang, Q., Zhuang, Z., Gu, S., Kaspar, R. B., Zheng, J., Wang, J., Qiu, S., Yan, Y. (2014). *Nat. Commun.*, **5**, 4503.

14. Waller, P. J., Lyle, S. J., Osborn Popp, T. M., Dierck, C. S., Reimer, J. A., Yaghi, O. M. (2016). *J. Am. Chem. Soc.*, **138**, 15519–15522.

15. Kandambath, S., Mallick, A., Likose, B., Mane, M. V., Heine, T., Banerjee, R. (2012). *J. Am. Chem. Soc.*, **134**, 19524–19527.

16. Rao, M. R., Fang, Y., Feyter, S. D., Perepichka, D. F. (2017). *J. Am. Chem. Soc.*, **139**, 2421–2427.

17. Dalapati, S., Jin, S., Gao, J., Xu, Y., Nagai, A., Jiang, D. (2013). *J. Am. Chem. Soc.*, **135**, 173101–17313.

18. Guo, J., Xu, Y., JIn, S., Gao, J., Chen, L., Kaji, T., Honsho, Y., Addicoat, M. A., Kim, J., Saeki, A., Ihee, H., Seki, S., Irle, S., Hiramoto, M., Gao, J., Jaing D. (2013). *Nat. Commnun.*, **4**, 2736.

19. Nagai, A., Chen, X., Feng, X., Ding, X., Guo, Z., Jiang, D. (2013). *Angew. Chem. Int. Ed.*, **52**, 3770–3774.

20. Shi, W., Xing, F., Bai, T.-L., Hu, M., Zhao, Y., Li, M-X., Zhu, S. (2015). *ACS Appl. Mater. Interface*, **7**, 14493–14500.

21. Kuhn, P., Antonietti, M., Thomas, A. (2008). *Angew. Chem. Int. Ed.*, **47**, 3450–3453.

22. Zhang, W., Qiu, L. G., Yuan, Y. P., Xie, A. J., Shen, Y. H., Zhu, J. F. (2012). *J. Hazard. Mater.*, **221**, 147–157.

23. Gutzler, R., Walch, H., Eder, G., Kloft, S., Heckl, W. H., Lackinger, M. (2009). *Chem. Commoun.*, **7**, 4456–4458.

24. Jin, E., Asada, M., Xu, Q., Dalapti, S., Addicoat, M. A., Brady, M. A., Xu, H., Nakamura, T., Heine, T., Chen, Q., Jiang, D. (2017). *Science*, **357**, 673–676.

25. Jackson, K. T., Reich, T. E., El-Kaderi, H. M. (2012). *Chem. Commun.*, **48**, 8823–8825.

26. Beaudoin, D., Maris, T., Wuest, J. D. (2013). *Nat. Chem.*, **5**, 830–834.

27. Höpfl, H. (1999). *J. Organomet. Chem.*, **581**, 129–149.

28. Sheepwash, E., Krampl, V., Scopelliti, R., Sereda, O., Neels, A., Sverin, K. (2011). *Angew. Chem. Int. Ed.*, **50**, 3034–3037.

29. Ding, S.-Y., Gao, J., Wang, Q., Zhang, Y., Song, C.-G., Su, C.-Y., Wang, W. (2011). *J. Am. Chem. Soc.*, **133**, 19816–19822.

30. Sakaushi, K., Antonietti, M. (2015). *Acc. Chem. Res.*, **48**, 1591–1600.

31. DeBlase, C. R., Silberstein, K. E., Truong, T.-T., Abruña, H. D., Dichtel, W. R. (2013). *J. Am. Chem. Soc.*, **135**, 16821–16824.

32. Kandambeth, S., Shinde, D. B., Panda, M. K., Lukose, B., Heine, T., Banerjee, R. (2013). *Angew. Chem. Int. Ed.*, **123**, 13052–13056.
33. Davis, M. S. (2002). *Nature*, **417**, 813–821.
34. Cundy, C. S., Cox, P. A. (2003). *Chem. Rev.*, **103**, 663–701.
35. Wan, Y., Zhao, D. (2007). *Chem. Rev.*, **107**, 2812–2860.
36. Sakamoto, Y., Kaneda, M., Terasai, O., Zhan, D. Y., Kim, J. M., Stucky, G. D., Shim, H. J., Ryoo, R. (2000). *Nature*, **408**, 449–453.
37. Che, S., Garcia-Bennett, A. E., Yokoi, T., Sakamoto, K., Kunieda, H., Terasaki, O., Tatsumi, T. (2003). *Nat. Mater.*, **2**, 801–805.
38. Shen, S. D., Garcia-Bennett, A. E., Liu, Z., Lu, Q. Y., Shi, Y. F., Yan, Y., Yu, C. Z., Liu, W. C., Cai, Y., Terasdaki, O., Zhao, D. Y. (2005). *J. Am. Chem. Soc.*, **127**, 6780–6787.
39. Tan, B., Dozier, A., Lehmler, H. J., Knutson, B. L., Rankin, S. E. (2004). *Langmuir*, **20**, 6981–6984.
40. Tranchemontague, D. J., Mendoza-Cortés, J. L., O'Keeffe, M., Yaghi, O. M. (2009). *Chem. Soc. Rev.*, **38**, 1257–1283.
41. Uemura, T., Yanai, N., Kitagawa, S. (2009). *Chem. Soc. Rev.*, **38**, 1228–1236.
42. Wang, Z., Cohen, S. M. (2009). *Chem. Soc. Rev.*, **38**, 1315–1329.
43. Shimizu, G. K. H., Vaidhyanathan, R., Taylor, J. M. (2009). *Chem. Soc. Rev.*, **38**, 1430–1449.
44. Li, J. R., Kuppler, R. J., Zhou, H. C. (2009). *Chem. Soc. Rev.*, **38**, 1477–1504.
45. Asefa, T., MAcLachlan, M. J., Coombs, N., Ozin, G. A. (1999). *Nature*, **402**, 867–871.
46. Eddaoudi, M., Kim, J., Rosi, N., Vodak, D., Wachter, J., O'Keeffe, M., Yaghi, O. M. (2002). *Science*, **295**, 469–472.
47. James, S. L. (2003). *Chem. Soc. Rev.*, **32**, 276–288.
48. Yaghi, O. M., O'Keefe, M., Ockwig, N. W., Chae, H. K., Eddoudi, M., Kim, J. (2003). *Nature*, **423**, 705–714.
49. Rowsell, J. L. C., Milward, A. R., Park, K. S., Yaghi, O. M. (2004). *J. Am. Chem. Soc.*, **126**, 5666–5667.
50. Chen, B. L., Liang, C. D., Yang, J., Contreas, D. S., Clancy, Y. L., Lobkovsky, E. B., Yaghi, O. M. (2006). *Angew. Chem. Int. Ed.*, **45**, 1390–1393.
51. Heibaum, M., Glorius, F., Escher, I. (2006). *Angew. Chem. Int. Ed.*, **45**, 4732–4762.

52. Kitagawa, S., Kitaura, R., Noro, S. (2004). *Angew. Chem. Int. Ed.*, **43**, 2334–2375.
53. Kepert, C. J., Prior, T. J., Rosseinsky, M. J. (2000). *J. Am. Chem. Soc.*, **122**, 2334–2375.
54. Sadakiyo, M., Yamada, T., Kitagawa, H. (2009). *J. Am. Chem. Soc.*, **131**, 9906–9907.
55. Chandler, B. N., Cramb, D. T., Shimizu, G. K. H. (2006). *J. Am. Chem. Soc.*, **128**, 10403–10412.
56. Zheng, M. H., Wang, B., Wang, X. Y., Chen, X. M., Gao, S. (2006). *Inorg. Chem.*, **128**, 10403–10412.
57. Agusti, G., Munoz, M. C., Gaspar, A. B., Real, J. A. (2009). *Inorg. Chem.*, **48**, 3371–3381.
58. Zhang, L. J., Yu, J. H., Xu, J. Q., Lu, J., Bie, H. Y., Zhang, X. (2005). *Chem. Commun.*, **8**, 638–642.
59. Campbell. N. L., Clowes, R., Ritchie, L. K., Cooper, A. I. (2009). *Chem. Mater.*, **21**, 204–206.
60. Dogru, M., Sonnauer, A., Gavryushin, A., Knochel, P., Bein, T. A. (2011). *Chem. Commun.*, **47**, 1707–1709.
61. Biswal, B. P., Chandra, S., Kandambeth, S., Lukose, B., Heine, T., Banerjee, R. (2013). *J. Am. Chem. Soc.*, **135**, 5328–5331.
62. Karak, S., Kandambeth, S., Biswal, B. P., Sasmal, H. S., Kumar, S., Pachfule, P., Banerjee, R. (2017). *J. Am. Chem. Soc.*, **139**, 1856–1862.
63. Ding, S. Y., Cui, X. H., Feng, J., Lu, G., Wang, W. (2017). *Chem. Commun.*, **53**, 11956–11959.
64. Marumoto, M., Dasari, R. R., Ji, W., Feriante, C. H., Parker, T. C., Marder, S. R., Dichtel, W. R. (2017). *J. Am. Chem. Soc.*, **139**, 4999–5002.
65. De La Hoz, A., Diaz-Ortiz, A., Moreno, A. (2005). *Chem. Soc. Rev.*, **34**, 164–178.
66. Ritchie, L. K., Trewin, A., Reguera-Galan, A., Hasell, T., Cooper, A. I. (2010). *Miroporous Mesoporous Mater.*, **132**, 132–136.
67. Friscic, T. (2012). *Chem. Soc. Rev.*, **41**, 3493–3510.
68. Friscic, T., James, S. L., Boldyreva, E. V., Bolm, C., Jones, W., Mack, J., Steed, J. W., Suslick, K. S. (2015). *Chem. Commun.*, **51**, 6248–6256.
69. James, S. L., Adams, C. L., Bolm, C., Braga, D., Collier, P., Priscic, T., Grepioni, F., Harris, K. D., Hyett, G., Jones, W. (2012). *Chem. Soc. Rev.*, **41**, 413–447.
70. Zhang, P., Dai, S. (2017). *J. Mater. Chem. A*, **5**, 16118–16127.

71. Das, G., Balaji Shinde, D., Kandambeth, S., Biswal, B. P., Banerjee, R. (2014). *Chem. Commnun.*, **50**, 12615–12618.

72. Chandra, S., Kandambeth, S., Biswal, B. P., Lukose, B., Kunjir, S. M., Chaudhary, M., Babarao, R., Heine, T., Banerjee, R. (2013). *J. Am. Chem. Soc.*, **135**, 17853–17861.

73. Troschke, E., Grätz, S., Lübken, T., Borchardt, L. (2017). *Angew. Chem. Int. Ed.*, **56**, 14149–14153.

74. Ding, S. Y., Wang, W. (2013). *Chem. Soc. Rev.*, **42**, 548–568.

75. De la Pena Ruigomez, A., Rodriguez-San-Miguel, D., Stylianou, K. C., Cavallini, M., Gentili, D., Liscio, F., Milita, S., Roscioni, O. M., Ruiz-Gonzalez, M. L., Carbonell, C. (2015). *Chemistry (Easton)*, **21**, 10666–10670.

76. Montoro, C., Rodriguez-San-Miguel, D., Polo, E., Escudero-Cid, R., Ruiz-Gonzalez, M. L., Navarro, J. A. R., Ocon, P., Zamora, F. (2017). *J. Am. Chem. Soc.*, **139**, 10079–10086.

77. Rodriguez-San-Miguel, D., Abrishamkar, A., Navarro, J. A., Rodriguez-Trujillo, R., Amabilino, D. B., Mas-Balleste, R., Zamora, F., Puigmarti-Luis, J. (2016). *Chem. Commun.*, **52**, 9212–9215.

78. Matsumoto, M., Dasari, R. R., Ji, W., Feriante, C. H., Parker, T. C., Marder, S. R., Dichtel, W. R. (2017). *J. Am. Chem. Soc.*, **139**, 4999–5002.

Chapter 2

Crystallization and Structural Linkages of COFs

COFs are prepared by integrating organic molecular building blocks into predetermined network structures entirely through covalent bonds. The crystallization problem has been conquered by DCC in synthesis and reticular chemistry in materials design. This chapter reviews recent progress in the crystallization and structural linkages of COFs.

2.1 b–O Linkages

With lightweight compositions and high crystallinity in view, boron-linked covalent organic framework (COFs) have been intensively studied, including their crystallization mechanism, design of new topologies, achievement of high surface areas, and their utilization in gas storage applications. The successful building of such crystalline materials gained from adequate reversibility of the b–O linkages, including boroxine ($-B_3O_3$), boronic ester ($-BO_2C_2$), and spiroborate ($-O_4B^-$), and containing borazine ($-B_3N_3$), for example.

2.1.1 Boroxine-Linked COFs

Boroxines are the cyclotrimeric anhydrides of boronic acids. Their properties and applications have been recently reviewed [1].

Covalent Organic Frameworks
Atsushi Nagai

ISBN 978-981-4800-87-7 (Hardcover), 978-1-003-00469-1 (eBook)
www.jennystanford.com

By virtue of boron's vacant orbital, boroxines are isoelectronic to benzene, but it is generally accepted that they possess little aromatic character [2]. Several theoretical and experimental studies have addressed the nature and structure of these derivatives [1]; in particular, the X-ray crystallographic analysis of triphenylboroxine has confirmed that it is virtually flat [3]. Boroxines are easily produced by the simple dehydration of boronic acids, either thermally through azeotropic removal of water or by exhaustive drying over sulfuric acid or phosphorous pentoxides [4]. These compounds can be employed invariably as substrates in many of the same synthetic transformations known to affect boronic acids. Interest in the applications of boroxines as end products has increased in the past decade. Their use has been proposed as flame retardants [5] and as functional materials [5]. The formation of boroxine cross-linkages has been employed as a means to immobilize blue light–emitting oligofluorene diboronic acids [6]. A study examined the thermodynamic parameters of boroxine formation in water (Eq. 2.1) [7]. Using hydrogen-1 nuclear magnetic resonance (1H NMR) spectroscopy, the reaction was found to be reversible at room temperature (RT) and the equilibrium constants, relatively small ones, were found to be subject to substituent effects. For example, boroxines with a para-electron-withdrawing group have smaller equilibrium constants. The observation was interpreted as an outcome of a back-reaction, that is, boroxine hydrolysis, that is facilitated by the increased electrophilicity of boron. Steric effects also come into play, as indicated by a smaller *K* value for orthotolyboronic acid compared to that for the paraisomer. Variable temperature studies have provided useful thermodynamic information, which was found consistent with a significant entropic drive for boroxine formation due to the release of three molecules of water.

$$3\ \mathrm{R{-}C_6H_4{-}B(OH)_2} \rightleftharpoons (\mathrm{R{-}C_6H_4BO})_3 + 3\ \mathrm{H_2O} \qquad (2.1)$$

$$K = \frac{[\text{boroxine}]\,[\mathrm{H_2O}]^3}{[\text{boronic acid}]^3}$$

According to the stereochemistry and thermodynamic information of boroxines, the emergence of COF chemistry was triggered by Yaghi and coworkers' seminal work on the self-condensation of aromatic polyboronic acids (Fig. 2.1) [8]. The successful crystallization of these materials (COF-1, COF-102, and COF-103) was attributed to the implementation of a closed reaction system to sustain the availability of water for maintaining reversible conditions conducive to crystal growth. The solvents used and their mixture are a handle to control the diffusion of building blocks into the crystallization mother liquor. The crystallization was accomplished intentionally slowly and terminated after sufficient time had elapsed.

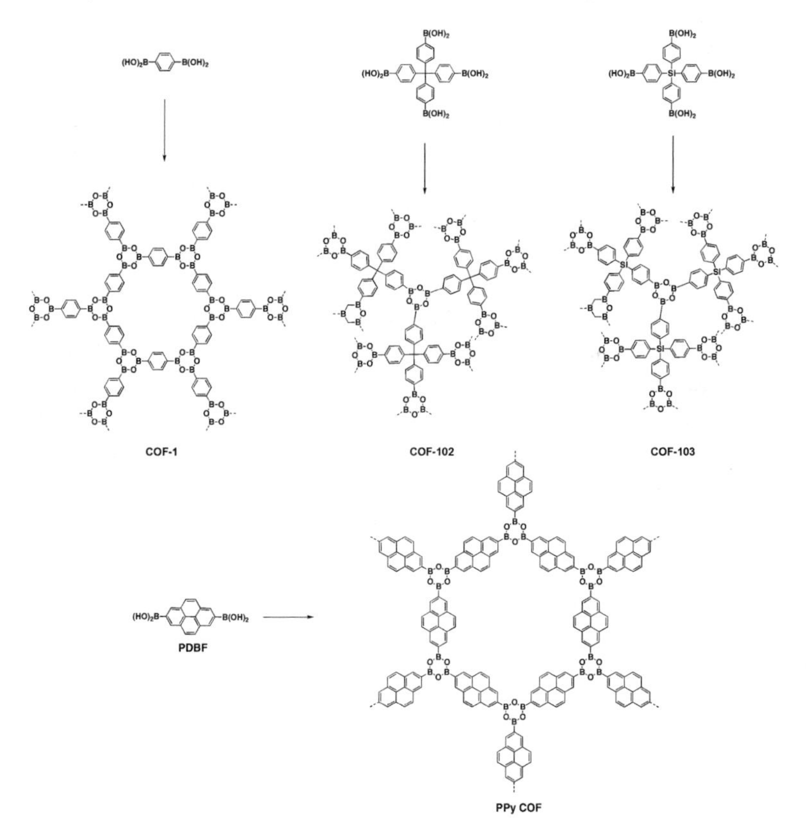

Figure 2.1 Construction of classic boroxine-linked COFs (COF-1, COF-102, COF-103) and pyrene-based boroxine-linked COF (PPy COF).

The selection of solvents and temperature to maximize the error correction associated with the boroxine ring–forming reactions can optimize the crystallinity of the boroxine-linked COFs. COF-1 with 2D architectures was prepared by the self-condensation of 1,4-phenylenediboronic acid under solvothermal conditions to produce an extended staggered layered structure with hexagonal pores (7 Å) and a Brunauer–Emmett–Teller (BET) surface of 711 m^2/g. Further, 3D COFs COF-102 and COF-103 were also synthesized by the self-condensation of tetrahedral (3D-Td) nodes tetra(4-dihydroxyborylphenyl)methane and its silane analog, respectively (Fig. 2.1) [9]. These two COFs are the crystal porous organic frameworks with high surface areas (3472 cm^2/g for COF-102 and 4120 m^2/g for COF-103) and a pore size distribution of 11.5 Å for COF-102 and 12.5 Å for COF-103.

After these pioneer reports, Jiang and coworkers demonstrated the first example of a photoconductive COF, in which sheets composed of arene building blocks lie one above the other in an eclipsed arrangement (Fig. 2.1) [10]. COF-1 possesses a pore 7 Å in diameter with a BET surface area of 711 m^2/g. On the other hand, a large π monomer, for example, pyrene-2,7-diboronic acid, under otherwise identical conditions yields PPy COF with a large pore size (18.8 Å) [10]. The large pyrene building blocks prefer a supericomposed π–π interaction. As a result, PPy COF has an eclipsed stack structure.

Microwave heating can accelerate the reaction times, yielding high amounts of much cleaner COFs. Cooper and coworkers reported the synthesis and purification of COF-5 and COF-105 by microwave heating. The microwave synthetic condition reported by them was 200 times faster than the solvothermal condition reported by Yaghi et al., without changing the physical properties of COFs [11].

2.1.2 Boronic Ester (Dioxaborole)-Linked COFs

$$\text{PhB(OH)}_2 + \text{C}_6\text{H}_4(\text{OH})_2 \xrightleftharpoons{-2\,H_2O} \text{PhB}(\text{O}_2\text{C}_6\text{H}_4) \qquad (2.2)$$

The preparation of boronic esters (dioxaboroles) from catechols is straightforward by generating water molecules under solvothermal conditions (Eq. 2.2). The overall process is an equilibrium, and the forward reaction is fast with preorganized catechols and particularly favorable when the boronate product is insoluble in the reaction solvent. Therefore, the cross-condensation of polyboronic acids with catechols opens the way to stitch different building blocks into one framework (Fig. 2.2). The maximum utilization of

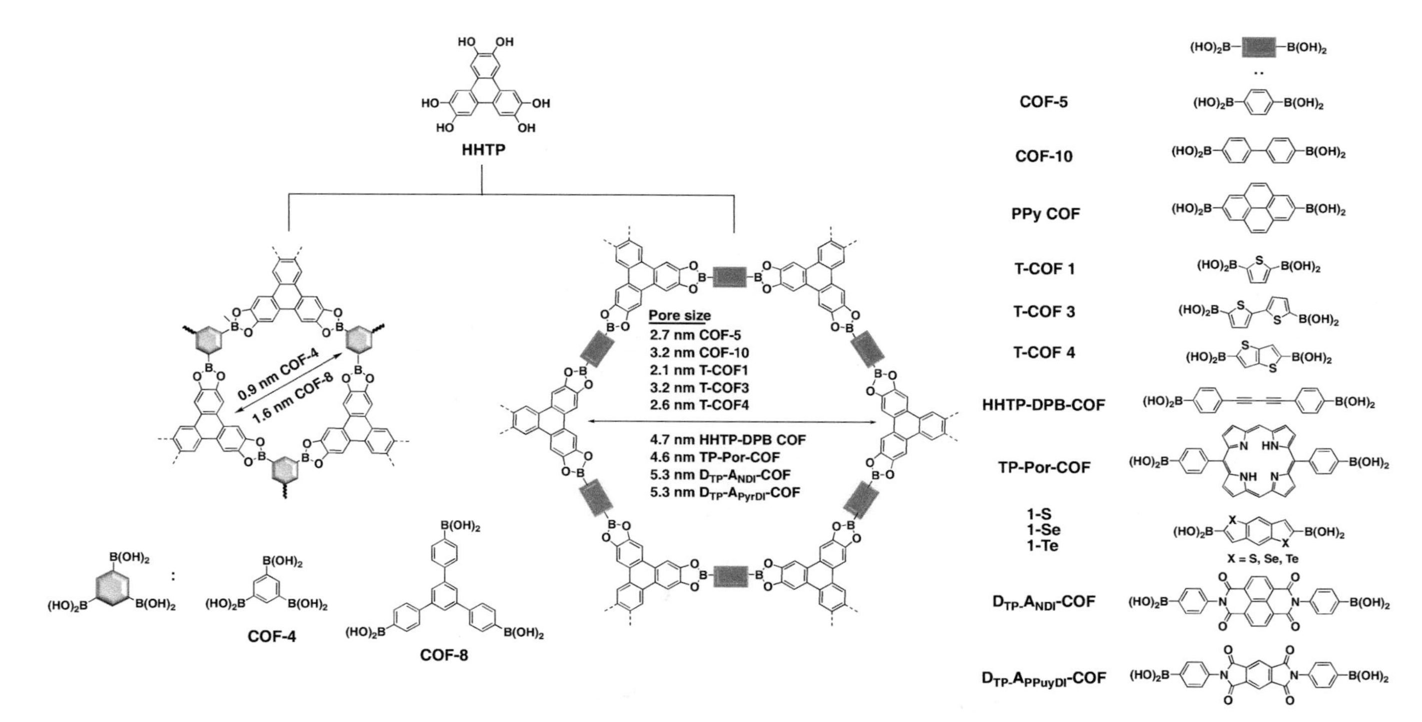

Figure 2.2 Construction of dioxaborole-linked COFs via the cross-condensation of triangular HHTP with ditopic or tritopic boronic acid to form 2D honeycomb layer structures.

-H_2O

THF/CH_3OH (99 : 1)
reflux for 3 days

R = H (18 Å)
= CH_3 (16 Å)
= CH_2CH_3 (14 Å)
= $CH_2CH_2CH_3$ (11 Å)

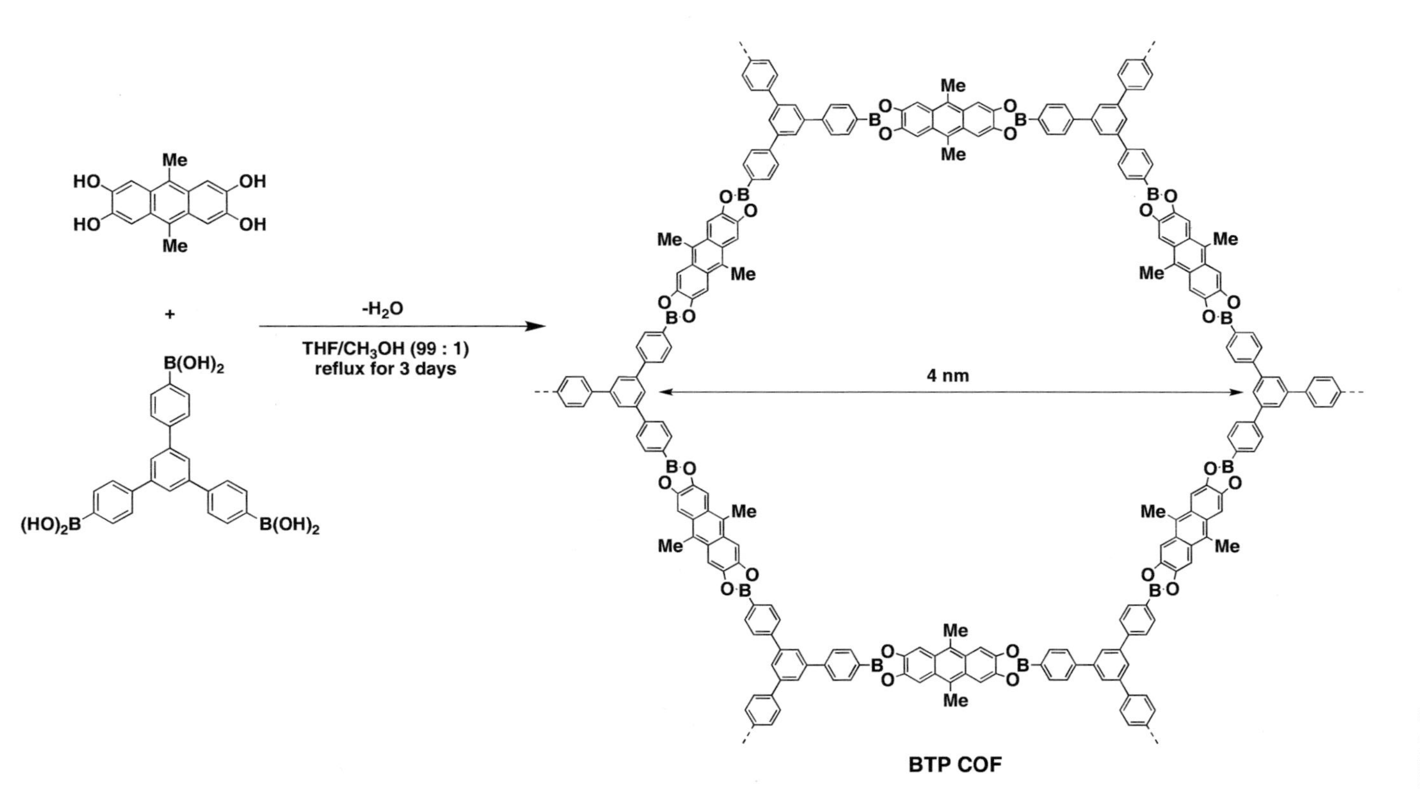

Figure 2.3 Construction of dioxaborole-linked COFs via the cross-condensation of ditopic catechol-based linkers with tritopic boronic acids to form 2D honeycomb layer structures. The pore sizes can be modulated by the decorated functionalities of the ditopic linkers and expanded by elongating both boronic acid and catechol linkers.

2,3,6,7,10,11-hexahydroxytriphenylene (HHTP) as a triangular building block with ditopic, tritopic, and tetratopic boronic acids leads to 2D and 3D networks (Fig. 2.2) [12–14]. Using a sealed reaction system under solvothermal conditions, COF-5, COF-6, COF-8, and COF-10 were successfully produced in the cocktail solvent system by mixing mesitylene and 1,4-dioxane for 3–5 days at 85°C–100°C [12, 14]. Therefore, the same COF using HHTP as the colinker can be also prepared by stirring under an inert atmosphere [15].

By using the tetrahydroxybenzene derivatives and 1,4-diboronic acid benzene, 2D hexagonal COFs can be prepared by stitching tritopic boronic acids together. The reaction was carried out in a mixed solvent of tetrahydrofuran and methanol (MeOH) by stirring under an insert atmosphere [16, 17].

Boronic ester linkage–built up COFs have high crystallinity but low hydrolytic and chemical stability due to the reversibility of the reactions, which leads to their decomposition upon exposure to water or acid. This disadvantage limits their applications. COF-anchored alkyl groups in the pores of COFs improve their stability by decreasing the hydrolysis rate of the connected linkage. Lavigne and coworkers were the first to incorporate alkyl groups into channel walls during the synthesis of boronate ester–linked COFs to increase the stability of COFs, as well as to fine-tune the pore sizes [17].

The presence of alkyl chains in the channels can help control the pore sizes so they remain in the range of 1.8 nm to 1.1 nm (Fig. 2.3), which was observed for the series of COFs with no alkyl groups (18 Å), COFs with methyl groups (16 Å), COFs with ethyl groups (14 Å), and COFs with propyl groups (11 Å). Meanwhile, the surface area decreased from 1263 to 105 cm^2/g and the pore volume decreased from 0.69 to 0.052 cm^3/g as the alkyl chain length increased. Interestingly, modification of the pore interior with increasingly larger alkyl groups causes a decline in nitrogen uptake but an increase in the molar amount of hydrogen adsorbed. Once they are submerged in aqueous media, the porosity of alkylated COFs decreases by about 25%, while the nonalkylated COFs are almost completely hydrolyzed, losing virtually all porosity. In addition, the degree of crystallinity decreases by about 40% for alkylated COFs and by 95% for nonalkylated COFs. By using microwave synthesis, Bein et al. extended the framework into materials having large (4 nm) openings (Fig. 2.3), featuring some of the largest pores in crystalline materials at this time [18].

Octahydroxyphthalocyanine (Pc) as a large π flat building block can also be utilized for the preparation of COFs. Dichtel's group was able to expand the pore size of the ZnPc lattice from 2.7 to 4.4 nm (Fig. 2.4) [19]. Furthermore, the selective and patterned growth of a 2D phthalocyanine COF on single-layer graphene was reported by Dichtel and coworkers. This provides a promising pathway to direct the growth of COF films [20].

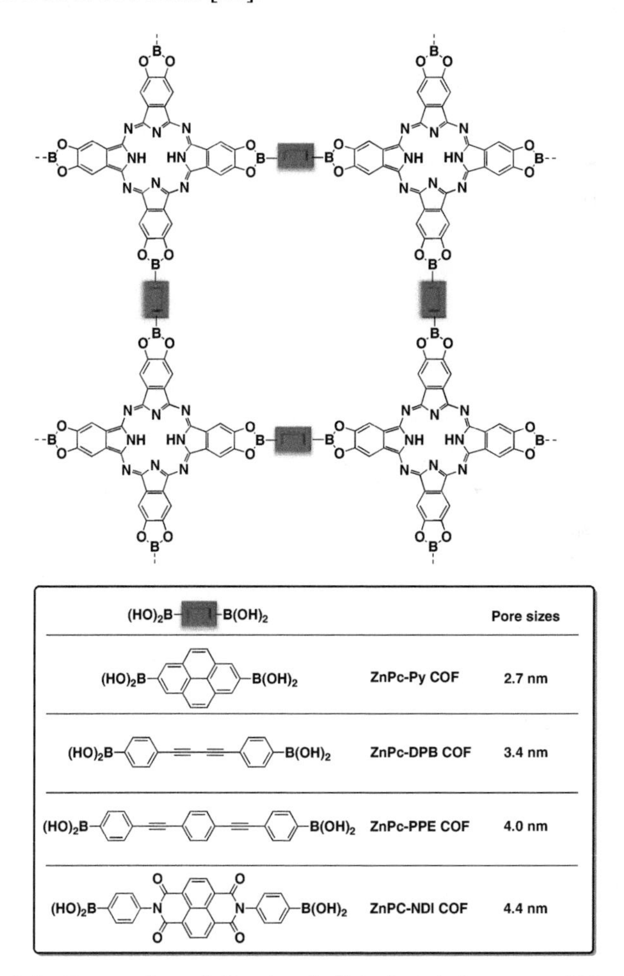

Figure 2.4 Preparation of dioxaborale-linked phthalocyanate COFs via the cross-condensation of ditopic catechol-based linkers with tritopic boronic acids to form a 2D honeycomb layer.

To avoid the oxidation and insolubility of polycatechols, protected catechols were proposed as starting building blocks for the synthesis of COFs, which could be deprotected in situ under catalysis by a Lewis acid such as trifluoro boron etholate ether complex ($BF_3{\cdot}OEt_2$) [21]. According to this strategy, triangular HHTP and square phthalocyanine as well as its metalated derivatives have been successfully incorporated into boronate ester–linked COFs (Fig. 2.5). Mechanistic studies present that before a Lewis acid is added, the boronic acids self-condensate reversibly to form boroxine-linked products and water. The addition of $BF_3{\cdot}OEt_2$ catalyzes acetonide hydrolysis of protected catechols, and the resulting catechol will rapidly condense with boronic acids to form boronate-linked COFs [22]. Once the free boronic acids are consumed, the formed products will hydrolyze and release more boronic acids, which will produce more boronate-linked COFs.

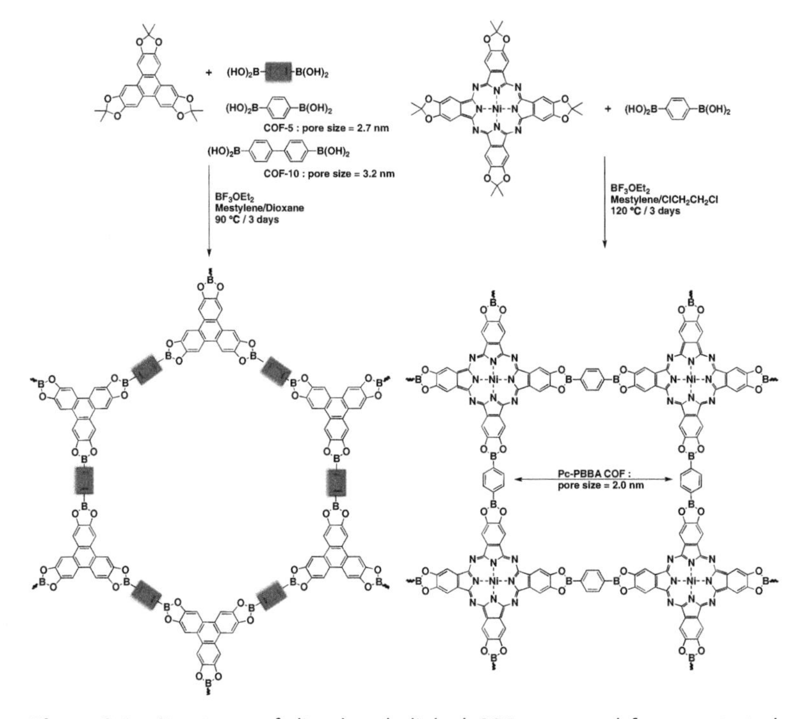

Figure 2.5 Structures of dioxaborole-linked COFs prepared from protected catechol-based linkers with boronic acids for improvement of solubility and antioxidation.

To focus on the preparation of highly crystalline COFs, a two-step microwave synthesis approach has been used for protected boronic acids as starting materials with the addition of HHTP. The resulting product proved to be highly crystalline and possessed large pore openings (Fig. 2.6). Control experiments did not yield COFs via the open-pot reaction. This work also provided the first high-resolution transmission electron microscopy image of a COF showing the pores clearly (Fig. 2.6b,c) [23].

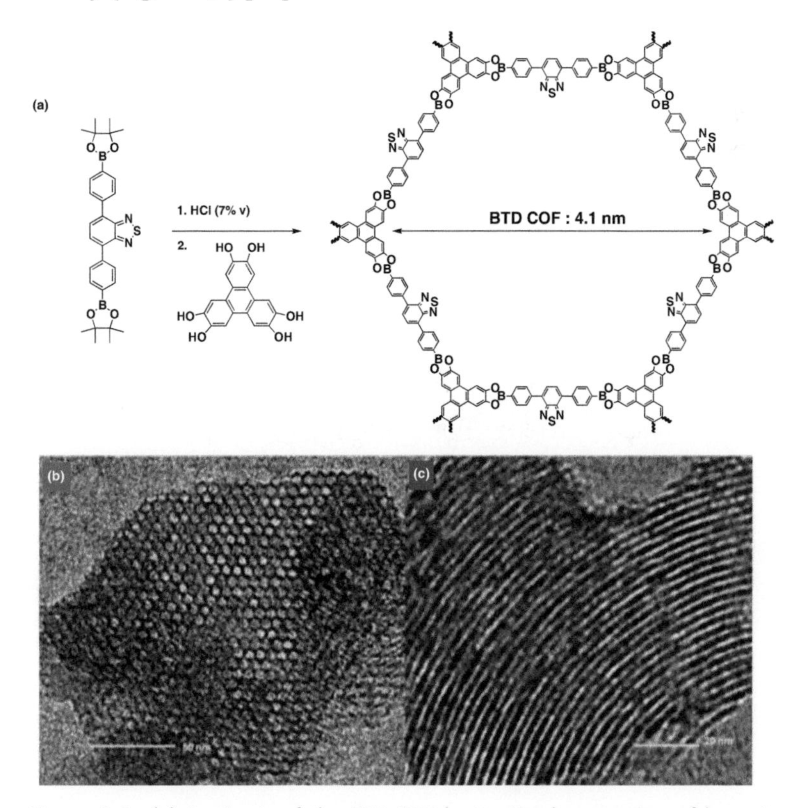

Figure 2.6 (a) Synthesis of the BTD COF by in situ deprotection of boronic pinacol ester, followed by the addition of HHTP via microwave heating, (b) projection along the columns showing the hexagonal pore structures, and (c) image of a crystal titled out of the columnar projection with a side view of the pores. Scale bar: (b) 50 nm and (c) 20 nm. Republished with permission of Royal Society of Chemistry, from Ref. [23], copyright (2013); permission conveyed through Copyright Clearance Center, Inc.

A modulation concept has been developed to control the structure and crystallinity of COFs, introducing a modulator in the preparation that can compete with one of the building blocks during solvothermal COF growth to form highly crystalline frameworks with a large domain size and very high porosity. Bein and coworkers utilized monoboronic acids as modulators in the solvothermal method of the archetypical COF-5 to optimize the crystallinity, domain size, and porosity of 2D COFs (Fig. 2.7). Furthermore, the

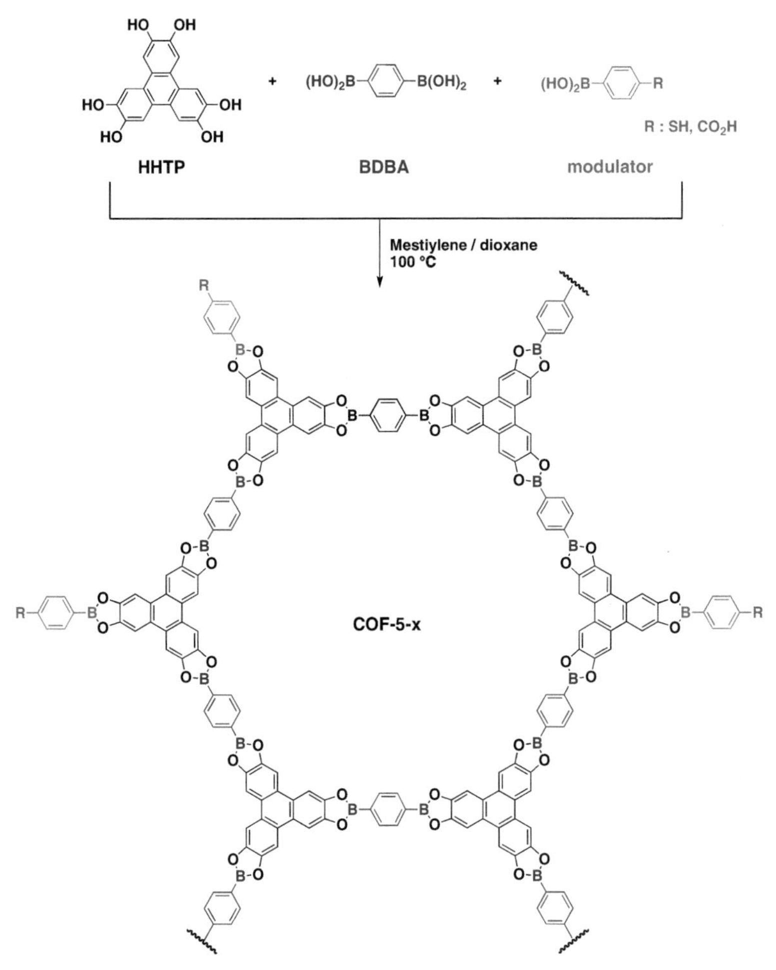

Figure 2.7 The modulation approach to prepare COF-5-x via the cross-condensation of HHTP and 1,4-boronic acid with terminal monoboronic acids with some functional groups to achieve higher crystallinity and external functionalities.

addition of monoboronic acids also provides for the potential construction of functional crystalline COFs [24]. The realization of highly crystalline COFs with the option of additional surface functionality will render the modulation concept beneficial for a range of applications, catalysis, and optoelectronics.

2.1.3 Spiroborate-Linked COFs

$$2\ \text{catechol (HO, HO)} \xrightleftharpoons{K_3B_4O_7\text{(base)}} \text{bis(catecholato)borate } B^- \cdot K^+ \qquad (2.3)$$

There have been many studies on reactions between boric acid and diols in connection with changes in conductivity, acidity, and rotatory polarization [25]. Hermans pointed out that the changes in these physical properties are due to the formation of a spiroborate complex that is produced from those reactions in a solution [26]. The first isolation of the spiroborate complex was performed by Böeseken and coworkers, who synthesized potassium biscatechol spiroborate from the reaction of catechol with potassium borate in water (Eq. 2.3) [25]. On the other hand, spiroborates are ionic derivatives of boronic acid, which have been reported to exhibit high resistance toward hydrolysis and stability in water, methanol, and under basic conditions [27, 28]. A spiroborate linkage can be formed readily through the condensation of polyols with alkali tetraborate [29, 30–32] or boric acid [33–35] or through the transesterification between borate and polyols [36] in a thermodynamic manner.

Recently, the condensation of diols with trialkyl borate in the presence of basic catalysts has been explored for the synthesis of spiroborate-linked COFs in which the negatively charged boron ions are located on the edges with different cations as counters ions. Zhang and coworkers constructed a novel type of spiroborate-linked ionic COF (ICOF, Fig. 2.8) that contains sp^3-hybridized boron anionic centers and tunable countercations (lithium or dimethylammonium) [37]. The prepared ICOFs (ICOF-1 and ICOF-2) show good thermal stabilities and excellent resistance to hydrolysis, remaining nearly intact when immersed in water or a basic solution for up to 2 days. A scanning electron microscope (SEM) image supports the single-crystalline morphology of both COFs. Powder X-ray diffraction (PXRD) results also show multiple sharp peaks, which indicate the orderliness of structures in the framework. However, the crystal structures of ICOFs have not been characterized. These ICOFs also have high BET surface areas (up to 1259 m^2/g) and adsorb a significant amount of H_2 (up to 3.11 wt%, 77 K, 1 bar) and CH_4 (up to 4.62 wt%, 273 K, 1 bar). The existence of permanently immobilized ion centers in ICOFs enables the transportation of lithium ions with RT lithium-ion conductivity of 3.05×10^{-5} S·L/cm and an average Li^+ transference number of 0.80 ± 0.02.

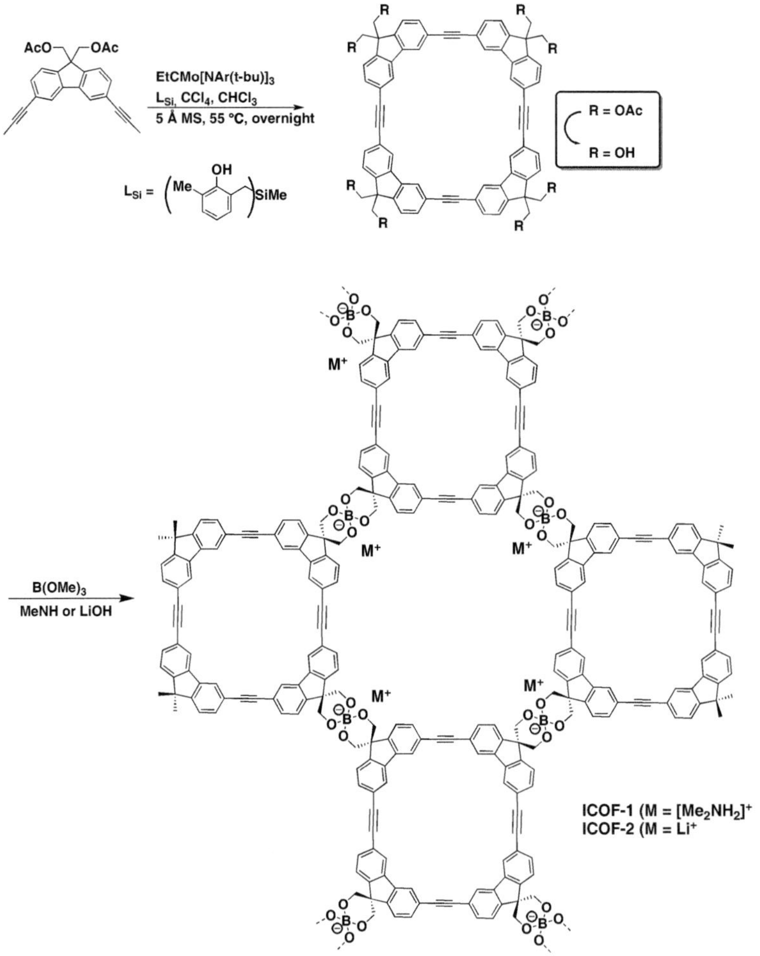

Figure 2.8 Construction of spiroborate-linked COFs featuring anionic skeletons with two counterions for high ionic conductivity.

More recently, Feng and coworkers have reported the first 3D anionic COFs based on flexible building blocks with different counterions, where γ-cyclodextrin (γ-CD) molecules act as organic struts that are covalently joined via spiroborate linkages. CD-COF-Li as a 3D anionic COF coordinated with Li^+ as a counterion was synthesized by the condensation of γ-CD and $B(OMe)_3$ in the presence of LiOH under microwave-assisted solvothermal conditions. The PXRD of the COF shows a highly crystalline structure and a BET surface area of 760 cm^2/g [37]. The microwave-assisted solvothermal synthesis ensures high production efficiency, good yields, and high purity [11]. When the proton acceptor in the reaction was changed to dimethylamine (DMA) or piperazine (PPZ), CD-COF-DMA and CD-COF-PPZ with the corresponding cations were obtained. According to structural analysis, all three CD COFs adopt an rra topology, where the nodes of the net are substituted with building blocks of the corresponding shape (Fig. 2.9). Owing to the high porosity, flexible building blocks, and charge skeleton, CD COFs show great potential in the fields of ion conduction and gas separation. The Li-ion conductivity of CD-COF-Li is as high as 2.7 mS/cm at 30°C, and this value is one of the highest Li-ion conductivities ever reported for crystalline porous materials, including COFs and metal-organic frameworks (MOFs).

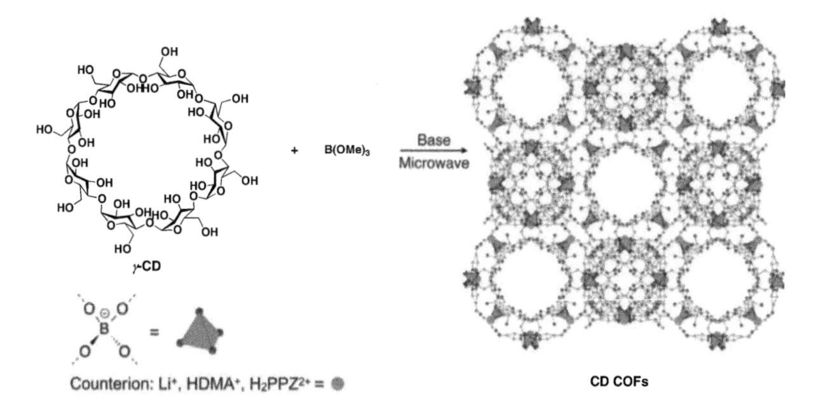

Figure 2.9 Construction of CD COFs condensed by γ-CD and $B(OMe)_3$ with LiOH, DMA, or PPz under microwave conditions. Reproduced with permission from Ref. [37]. Copyright (2017), John Wiley and Sons.

2.1.4 Borazine-Linked COFs

$$3\ \mathrm{C_6H_5{-}NH_2{-}BH_3} \xrightarrow{-H_6} (\mathrm{C_6H_5})_3\mathrm{N_3B_3H_3} \tag{2.4}$$

Borazine ($H_3B_3N_3H_3$), isolated by Stock and Pohland in 1926, is often dubbed as "inorganic benzene" due to its similarities with benzene: they are both liquid at RT, show equalized bond lengths (1.40 Å for benzene and 1.44 Å for borazine, with the latter being between B–N [single bond] at 1.51 and B=N [double bond] at 1.31 Å) and share a planar hexagonal structure. However, borazine shows only a weakly aromatic character and displays a great tendency to undergo hydrolysis to form boric acid and ammonia in the presence of moisture. Another difference is the string polar character of the B–N bonds resulting from the electron donation of the nitrogen atoms to the electrophilic boron centers.

Previously, Sneddon et al. obtained boron nitride by pyrolysis of polyborazylene polymer that had been obtained by thermal polymerization of borazine $B_3N_3H_6$ [38]. N- or B-substituted borazines are also interesting compounds that may also be polymerized. These polymers, via pyrolysis at high temperatures, could lead to ceramic-like boron nitride and boron carbonitride [38, 39]. *N*-substituted borazines are accessible by the thermolysis of primary amine-borane complexes $RNH_2{\cdot}BH_3$, usually prepared by the reaction of lithium borohydride with primary amine salts [40]. Manners and coworkers prepared borazines from $NH_3{\cdot}BH_3$ or $CH_3NH_2{\cdot}BH_3$ at a low temperature (45°C) [41]. Kinetic studies showed that the rate-determining step for both substrates is the loss of the last H2 molecule, which have been demonstrated to be fast only at high temperatures [42], leading to the dynamic covalent chemistry (DCC) concept. As shown in Eq. 2.4, a 99% yield of tri-*N*-phenylborazine could be obtained from $PhNH_2{\cdot}BH_3$ only after 30 min. at 120°C.

Borazine has been mainly used for the fabrication of BN-based ceramics or in organic optoelectronics [43]. However, to date, the use of borazine as a building block for the preparation of porous polymers remains scarce. El-Kaderi and coworkers first reported the targeted synthesis of the crystalline borazine-linked COF [BLP-2(H)] (Fig. 2.10) [44] and investigated its structural aspects, porosity, and performance in hydrogen storage. BLP-2(H) was prepared by the thermal decomposition of 1,3,5-(*p*-aminophenyl)-benzene-borane in a solvent mixture of mesitylene-toluene at 120°C/150 mTorr in a sealed tube for 3 days, which gave a good yield of BLP-2(H) as a white microcrystalline powder. The BET surface area and pore size of this COF was 1178 m^2/g and 6.4 Å, respectively. Using the Langmuir model (P/P_0 = 0.05–0.30), the surface area value was determined to be 1564 m^2/g. For comparison, the Langmuir surface area value of BLP-2(H) is similar to the Connolly surface area predicted for the eclipsed model (1840 m^2/g) and is much lower than that of the staggered model (3377 m^2/g), which further supports the formation of AA-eclipsed stacking. It can store up to 2.4 wt% of hydrogen at 77 K and 15 bar with isosteric heat of adsorption of 6.8 kJ/mol.

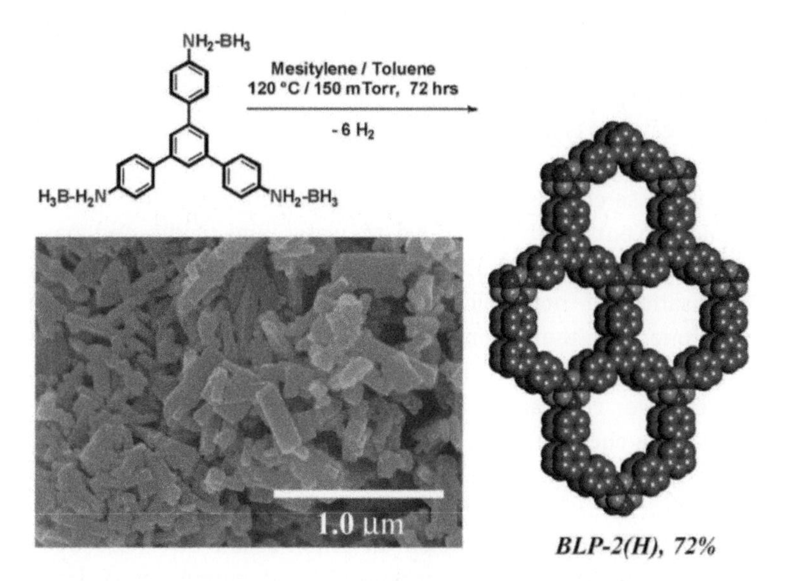

Figure 2.10 Construction of borazine-linked COF [BLP-2(H)] decomposed by 1,3,5-(*p*-aminophenyl)-benzene-borane and SEM image of as-prepared materials (inset). Reproduced with permission of Royal Society of Chemistry, from Ref. [44], copyright (2012); permission conveyed through Copyright Clearance Center, Inc.

2.2 Imine Linkages

$$R\text{-}NH_2 + R'\text{-}C(=O)\text{-}H \rightleftharpoons R\text{-}N{=}CH\text{-}R' + H_2O \quad (2.5)$$

Nowadays, Schiff base chemistry or dynamic imine chemistry is profusely utilized for the synthesis of COFs. Imine is a functional group or chemical compound containing a carbon-nitrogen double bond and is typically prepared by the condensation of primary amines and aldehydes and less commonly ketones (Eq. 2.5). In terms of mechanism, such reactions proceed via nucleophilic addition, giving a hemiaminal –C(OH)(NHR)– intermediate, followed by the elimination of water to give the imine (–N=C–) compound. The equilibrium in this reaction usually favors the carbonyl compound and amine, so azeotropic distillation or use of a dehydrating agent, such as molecular sieves or magnesium sulfate, is required to push the reaction in favor of imine formation. In recent years, several reagents, such as tris(2,2,2-trifluoroethyl)borate, pyrrolidine, acetic acid, and titanium ethoxide, have been shown to catalyze imine formation. Especially, according to synthesis of COFs, aqueous acetic acid as a catalyst is most useful.

Imine-based COFs constitute the largest amount of COFs based on Schiff base chemistry. In general, they are quite stable in most organic solvents and insensitive to water and acidic and basic conditions. Imine-based COFs were discovered by a reversible condensation of polytopic anilines with polybenzaldehydes or polyketone, with the elimination of water by a catalyst of aqueous acetic acid (Fig. 2.11). Generally, the crystallinity of the imine-based COFs is lower compared with that of boronate ester-linked COFs whereas the chemical stability in the presence of water, acids, and bases is significantly enhanced.

The first imine-based COF was developed by Yaghi and coworkers in 2009 [45]. They constructed an imine-based COF (COF-300), which shows a 3D fivefold interpenetrating diamond-like skeleton. COF-300 has a BET surface area of 1360 m^2/g and a pore size of 7.8 Å, which provides better hydrolytic stability compared to the boron-containing COFs. Furthermore, the broad range of multifunctional amines and aldehydes gives a large number of structural possibilities for imine-based COFs.

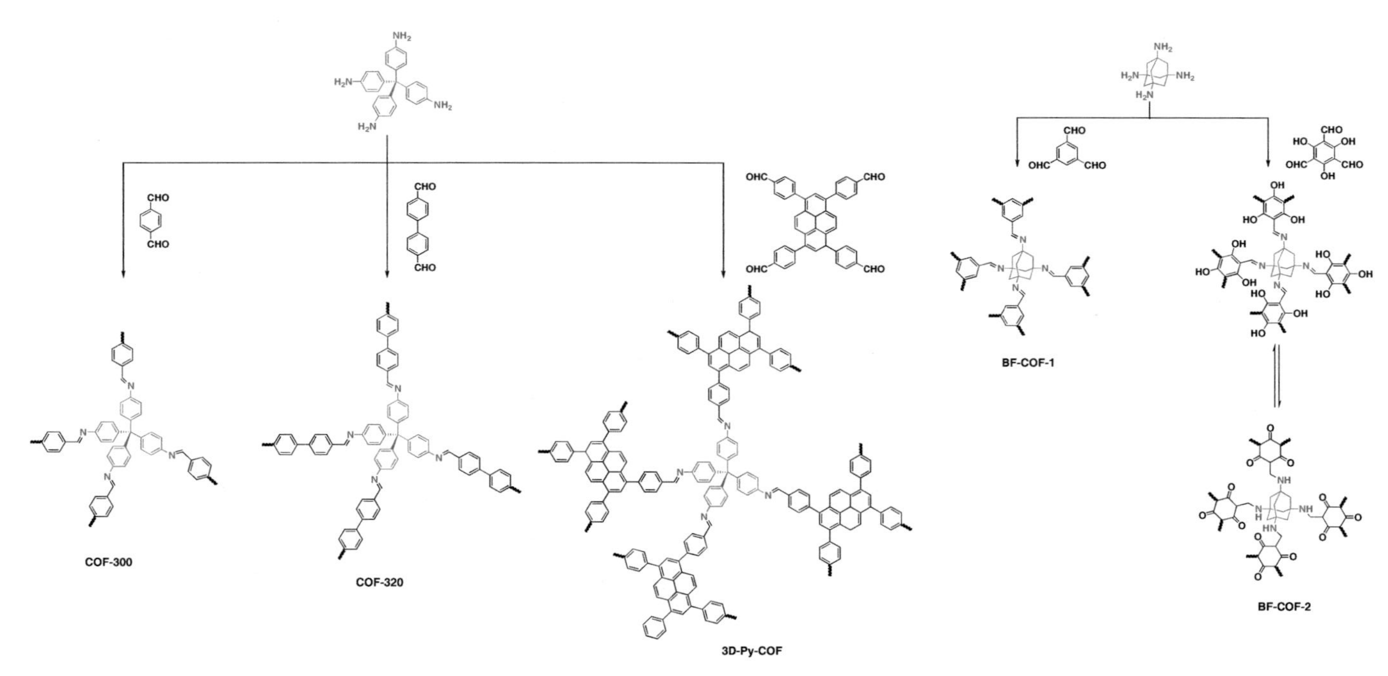

Figure 2.11 Construction of imine-linked COFs by co-condensation of tetratopic anilines—tetra-(4-anilyl)methane and 1,3,5,7-tetraaminoadamantane—with various ditopic, tetraaldehydes, and triangular aldehyde to form 3D networks.

Imine-based COF-320 was prepared via solvothermal condensationoftetra(4-anilyl)-methaneand4,4′-bisphneyldialdhyde in 1,4-dioxane at 120°C. A ninefold interlaced diamond network was formed, and the final crystalline material presented high porosity with a Langmuir surface area of 2400 m^2/g. The crystal structure of the COF-320 solid was elucidated by single-crystal 3D electron diffraction [46]. Recently, a pyridyl-functionalized version of COF-320 was synthesized, namely LZU-301, in which the reversible dynamic response upon guest accommodation and release was uncovered. Thus, the design principle of 3D dynamic COFs was claimed [47].

Wang and coworkers developed two new 3D pyrene-based COFs (3D-Py-COF) by the condensation of tetra(*p*-aminophenyl)methane and 1,3,6,8-tetrakis(4-formylphenyl)pyrene. This topology of a 3D Py COF is pts, which was first reported in this work. To be built up by pyrene in the 3D frameworks, this COF shows fluorescent properties and can be used in the detection of explosives [48].

Novel 3D microporous functionalized COFs (BF-COF-1 and BF-COF-2) were designed and synthesized by Yan and coworkers from the reaction of a tetrahedral alkyl amine, 1,3,5,7-tetraaminoadamantane, and 1,3,5-triformylbenzene (TFB) or triformylphloroglucinol [49]. These BF COFs were used for Knoevenagel condensation reaction with high conversion (BF-COF-1, 96%; BF-COF-2, 98%), highly efficient size selectivity, and good recyclability. Adamantane as the rigid building group is the key for forming targeted products. In contrast, the combination of tetra(*p*-aminophenyl) with TFB could not produce a crystalline product.

A new strategy to enhance both the chemical stability and crystallinity in 2D porphyrin COFs was proposed by Banerjee [50] and Jiang [51] groups. In this method, the imine bond (–C=N–) in the COF interior frame is protected by introducing –OH to the Schiff base (–C=N) centers in COFs and creating an intramolecular [–O–H···N=C–] hydrogen bond (Fig. 2.12) [50]. This hydrogen-bonding interaction enhances the stability of imine bonds in the presence of water and acid (3 N HCl). Otherwise, this hydrogen bond in 2,3-dihydroxyterephthalaldehyde–5,10,15,20-tetrakis(4-aminophenyl)-21*H*,23*H*-porphyrin (DhaTph) also could enhance the crystallinity and porosity compared to the methoxy-substituted COF DmaTph, in which this intramolecular hydrogen bond is absent.

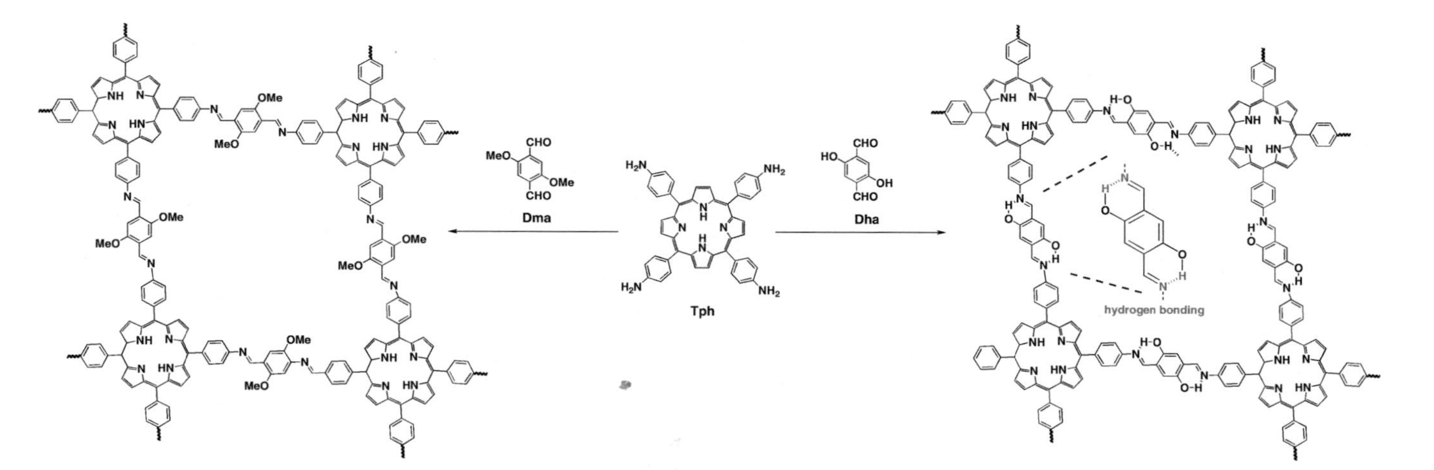

Figure 2.12 Synthesis of DmaTph and DhaTph by the condensation of square planar Tph building unit and linear Dma and Dha building units.

The formation of a crystalline structure via relatively weak interactions between planar layers has been well defined and designed in crystal engineering. Jiang and coworkers prepared a series of 2D COFs locked with intralayer hydrogen-bonding interactions (Fig. 2.13) [51]. The hydrogen-bonding interaction sites were located on the edge units of the imine-linked tetragonal porphyrin COFs, and the contents of the hydrogen-bonding sites in the COFs were synthetically regulated by using a three-component condensation system. The intralayer hydrogen-bonding interactions suppress the torsion of the edge units and lock the tetragonal sheets in a planar conformation. This planarization enhances the interlayer interactions and triggers extended π-cloud delocalization over 2D sheets. All pores in the COFs are 2.5 nm in size. These COFs have an AA stacking layered structure. As a result, COFs with layered 2D sheets amplify these effects and strongly affect the physical properties of the materials, including improving their crystallinity, enhancing their porosity, increasing their light-harvesting capability, and reducing their bandgap.

In addition to high stability and crystallinity of COFs by hydrogen-bonding interaction, self-complementary π electronic forces provide new opportunities for enhancing the crystallinity of COFs. Jiang and Nagai groups reported a synthetic method to control the crystallinity and porosity of COFs by managing interlayer interactions based on self-complementary π electronic forces (Fig. 2.14).

A three-component condensation reaction of CuP and 1,4-diformylbenzene with 2,3,5,6-tetrafluoroterephthalaldehyde can control the crystal structure of the CuP-Ph COF. Fluoro-substituted aromatic units in different ratios were integrated into the edge units, which will induce the self-complementary π electronic interactions in the COFs.

This interaction increases the crystallinity and porosity of COFs by maximizing the total crystal stacking energy and minimizing the unit cell size. This work provides a new pathway to improve the crystals of COFs by controlling the interlayer interactions [52]. The same authors have also developed a nice strategy to soften the polarization influence of the C=N bond on the destabilization of the layered structure in hexagonal 2D COFs based on imine reactions [53]. In imine-linked COFs, the C=N bond is polarized to yield partially positively charged carbon and negatively charged

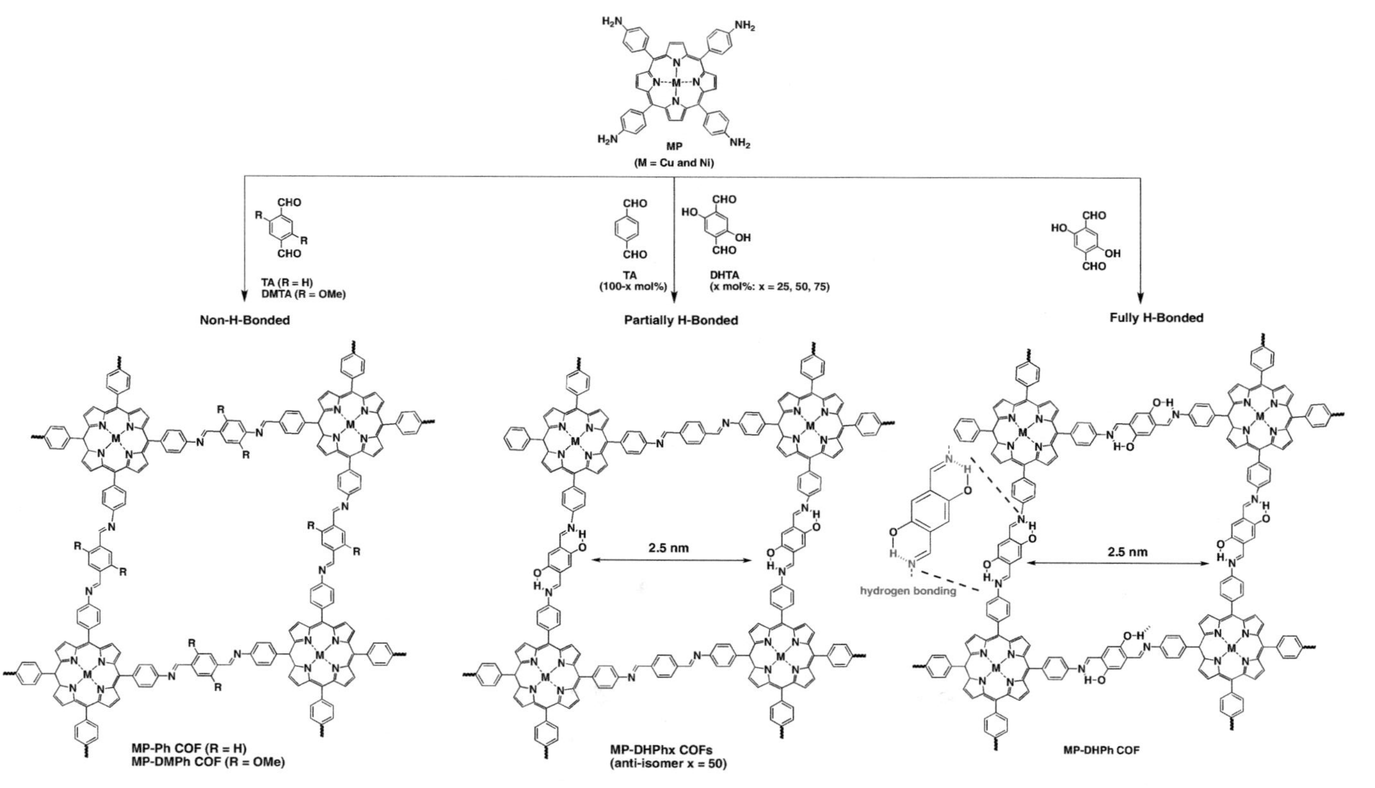

Figure 2.13 Construction of 2D porphyrin COFs with designable content of hydrogen-bonding structures.

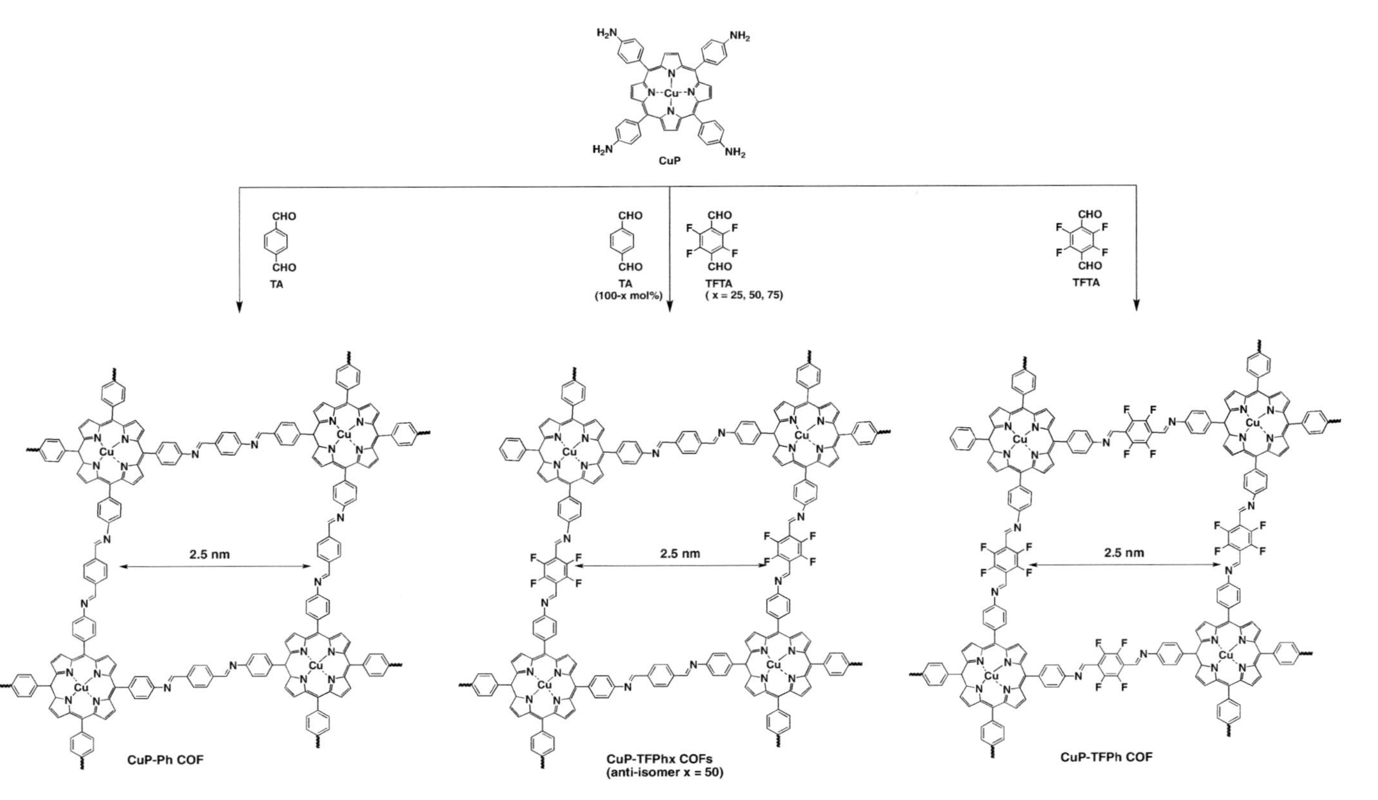

Figure 2.14 Construction of 2D COFs integrated with self-complementary π electronic forces (CuP-TFPh$_X$, where X = 25, 50, and 75 mol%) and CuP-Ph and CuP-TFPh controls.

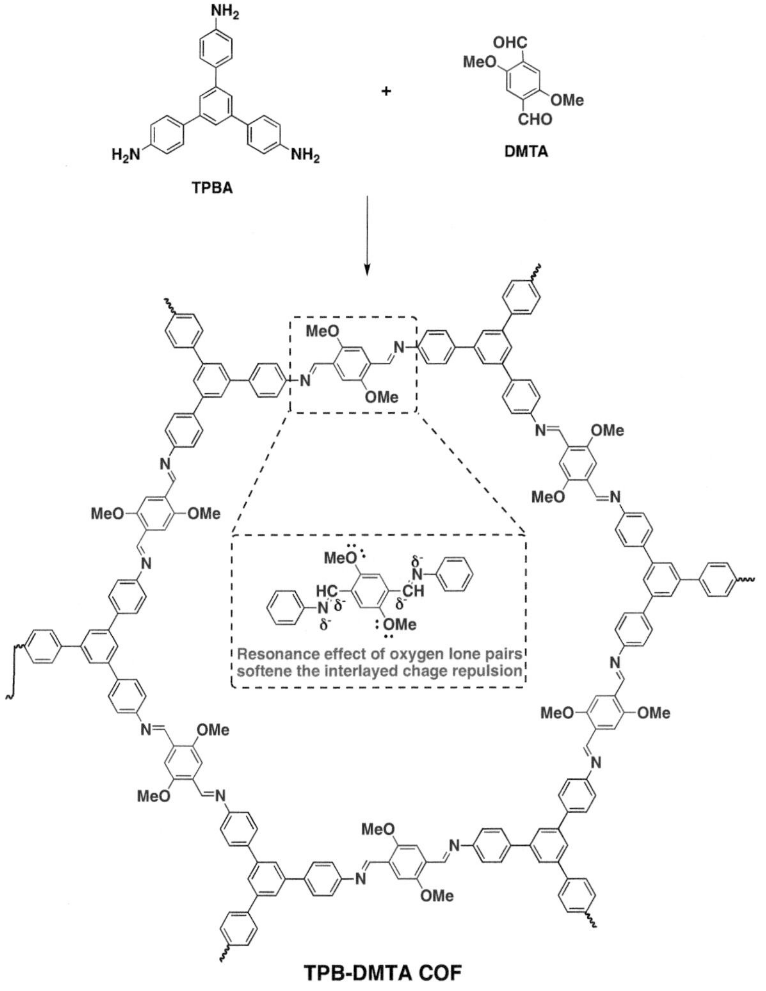

Figure 2.15 Synthesis of a TPB-DMTP COF via the condensation of DMTA and TPBA. Inset: the structure of the edge units of the COF and the resonance effect of oxygen lone pairs that weaken the polarization of the C=N bonds and soften the interlayer repulsion in the COF.

nitrogen. Thus, in a hexagonal 2D COF, each macrocycle consists of 12 polarized C=N segments; the aggregation of a large number of charge groups causes electrostatic repulsion and destabilizes the layered structure, as predicted theoretically [54]. Therefore, in this case, crystallinity and stability were enhanced by incorporating methoxy groups in the pore walls of COFs. Introducing the two electron-donating methoxy groups to each phenyl edge delocalizes the two lone pairs from the oxygen atoms over the central phenyl ring, which reinforces the interlayer interactions and so stabilizes the COF and aids in its crystallization (Fig. 2.15).

On the other hand, the utilization of macrocyclic host derivatives with a shallow rigid cavity as triformylcyclotrianisylene, a derivative of the well-known cyclotrianisylene (CTV), seems to stabilize the perfect eclipsed stack model rather than COFs based on planar motifs. The columnar stacking of crown CTV motifs avoids sliding between layers in CTV COFs (Fig. 2.16) [55].

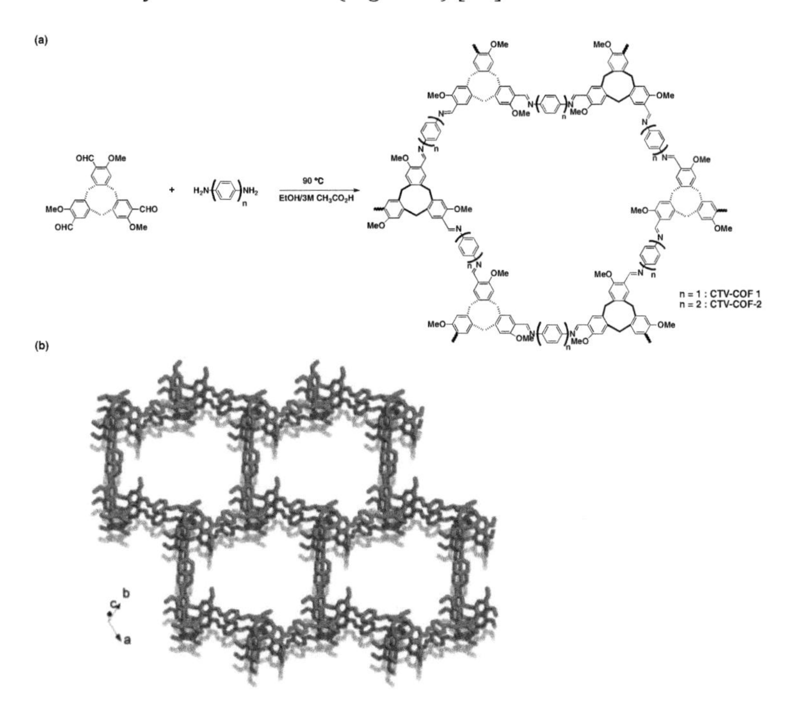

Figure 2.16 (a) Synthetic route of CTV COFs and (b) stick view of CTV COF-1. All hydrogen atoms are omitted for clarity. Different layers are shown as different colors. Adapted with permission of Royal Society of Chemistry, from Ref. [55], copyright (2014); permission conveyed through Copyright Clearance Center, Inc.

Regarding the size and shape of the pores in 2D COFs, it is known that they can be well tailored by changing the structures of monomers [16, 22, 56, 57]. In general, COFs with hexagonal and tetragonal topologies are by far the most extensively investigated. However, modifications of these topologies can introduce structural changes in the COFs. Hence, COFs with a trigonal topology have been rationally designed by using C_6-symmetrical vertices as hexaphenylbenzene (HPB) and hexabenzocoronene, thus providing small pore sizes and high π column densities [58].

To achieve remarkable chemical stability, Banerjee and coworkers explored a new method to enhance the chemical stability of imine COFs by the transformation of combined reversible Schiff base bonds to irreversible b-ketoenanime bonds (Fig. 2.17) [59]. They synthesized the COFs TpPa-1 and TpPa-2 by the Schiff base reactions of 1,3,5-triformylphloroglucinol (Tp) with p-phenylenediamine (Pa-1) and 2,5-dimenthyl-p-phenylenediamine (Pa-2), respectively. The expected enol-imine (OH) form underwent irreversible proton tautomerization into irreversible ketoenamine form, which confers outstanding stability to boiling water, aqueous acid (9 N HCl), and base (9 N NaOH). This linkage has also been synthesized by a simple solvent-free RT mechanochemical synthetic route [60]. The amide-linked COFs show improved chemical stability relative to their imine progenitors but a low degree of crystallinity. Recently, Yaghi and coworkers developed a method for the chemical conversion of an imine bond to an amine bond inside COFs to improve both chemical stability (including base/acid stability) and crystallinity (Fig. 2.18). They converted two layered imine COFs to amide-linked COFs (4PE-1P-COF 1; 1′ and 4PE-1P-COF 1′; 2′) without loss of their crystallinity or topology. Fourier-transform infrared spectroscopy and 13C CP-MAS NMR spectra indicated the successful conversion of COF-1 and COF-2 to COF-1′ and COF-2′. The high crystal structures of COF-1′ and COF-2′ were examined by X-ray diffraction analysis. The BET surface areas of COF-1′ and COF-2′ were compared to those of COF-1 and COF-2 to examine the increase in framework mass and decrease in pore volumes before and after the reduction. This method offers a new pathway to overcome the usual crystallization problem in COF chemistry [61].

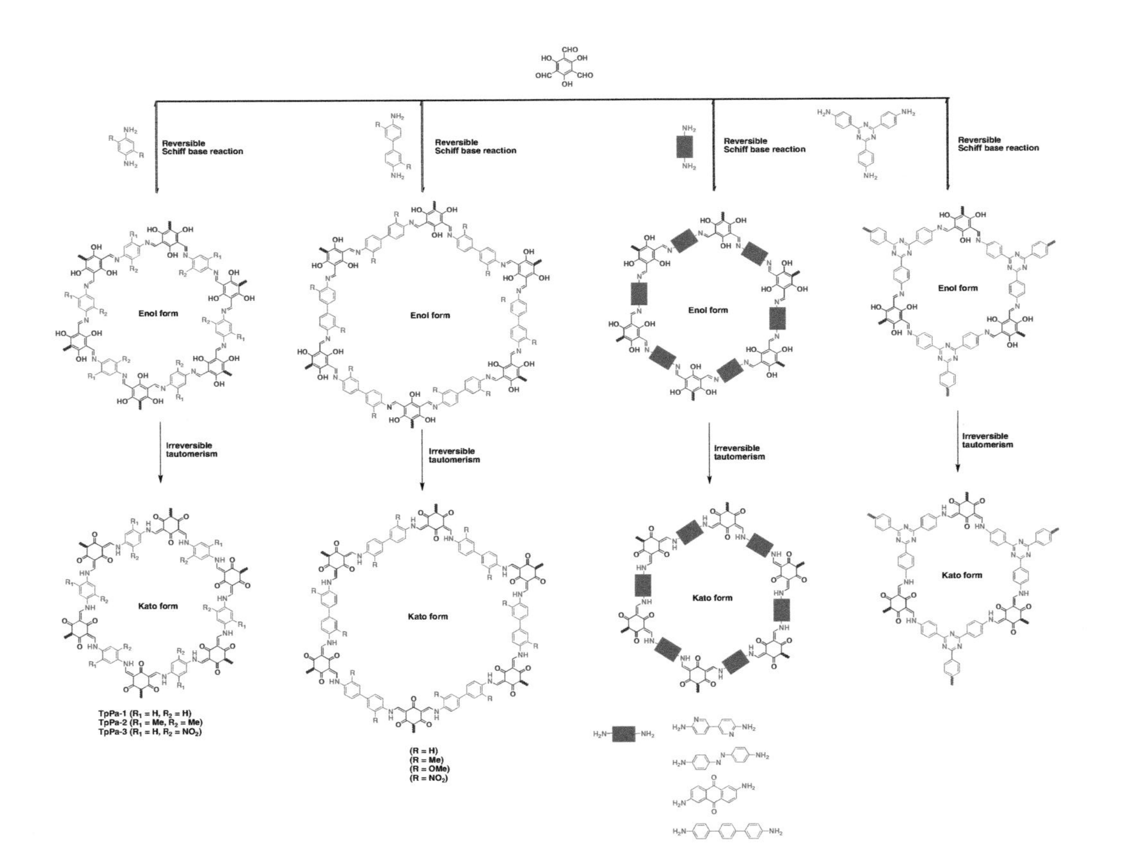

Figure 2.17 Synthesis of secondary amine-linked COFs through tautomerism by linking Tp with ditopic or tritopic aniline-based COFs.

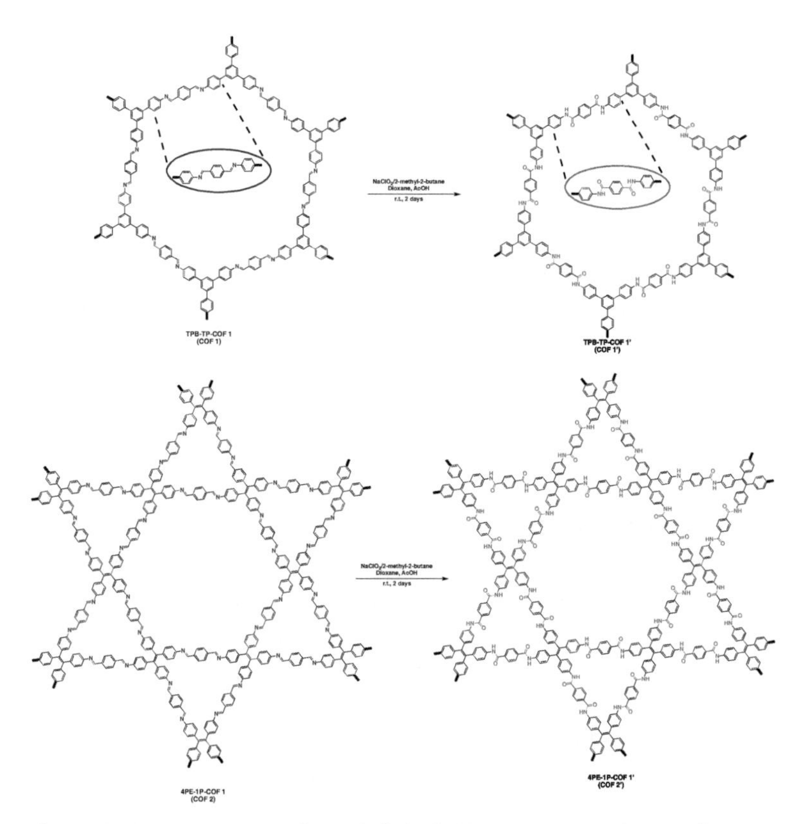

Figure 2.18 Preparation of amide-linked COFs via postoxidation of imine-linked COFs, which bypass the difficulty of crystallization for irreversible amide bonds.

2.3 Hydrazone Linkages

$$R_1C(=O)R_2 + \underset{\text{hydrazine}}{H_2N\text{-}NH_2} \rightleftharpoons R_1R_2C{=}N\text{-}NH_2 + H_2O \quad (2.6)$$

Hydrazones are valuable and versatile building blocks in synthetic chemistry [62]. Owing to their similarities to carbonyl compounds, hydrazones are a class of organic compounds with the structure $R_1R_2C{=}NHN_2$. They are related to ketones and aldehydes by the replacement of the oxygen with the NH_2 functional group. They are formed usually by the action of hydrazine on ketones or aldehydes, as shown in Eq. 2.6.

Like the reaction of imine-linked COFs, the reaction of benzaldehyde and hydrazide groups could yield hydrazine-linked COFs (Fig. 2.19). Hydrazone-linked COFs represent good chemical stability relative to imine-linked ones because of the hydrogen-bonding interaction between the oxygen atom in the alkoxyl chain and the hydrogens in amide (–CONH–) units. However, the functional monomers available for the condensation of hydrazone-based COFs are limited. Recently, the first hydrazone-based COFs were developed by the condensation reaction of a difunctional acylhydrazide (2,5-diethoxyterephthalohydraide) and trifunctional aldehydes (1,3,5-triformylbenzene or 1,3,5-tris(4-formylphenyl) benzene) to provide two crystalline mesoporous COFs, COF-42 and COF-43 [63].

The exfoliation of COF-43 in various solvents was done by Dichtel and coworkers to create few-layered 2D polymers. The obtained exfoliated COF-43 showed by PXRD an apparent loss of crystallinity but no apparent changes in its covalent linkages and high aspect ratios [64].

Stegbauer et al. reported a hydrazone-based COF—1,3,5-tris(4-formylphenyl)triazine (TFTP) COF—capable of visible-light-driven hydrogen generation with Pt as a proton reduction catalyst. The COF, which was constructed by TFPT and 2,5-diethoxy-terephthalohydrazide building blocks, shows a layered structure with a honeycomb-type lattice featuring mesopores of 3.8 nm and a BET surface area of 1603 m^2/g. On irradiation by visible light, the Pt-doped COF continuously produces hydrogen from water, without signs of degradation. This new application of COFs in photocatalysis opens new pathways for heterogeneous catalysts [65].

A thioether-based hydrazone-linked COF (COF-LZU8) was designed rationally by Ding and coworkers, and they used it for highly sensitive detection and effective removal of Hg^{2+}. Regarding the high stability of the hydrazone linkages, together with a dense distribution of the thioether groups and the straight channels in COF-LZU8, the recycling of COF-LZU8 achieved the simultaneous detection and removal of the toxic Hg^{2+} [66].

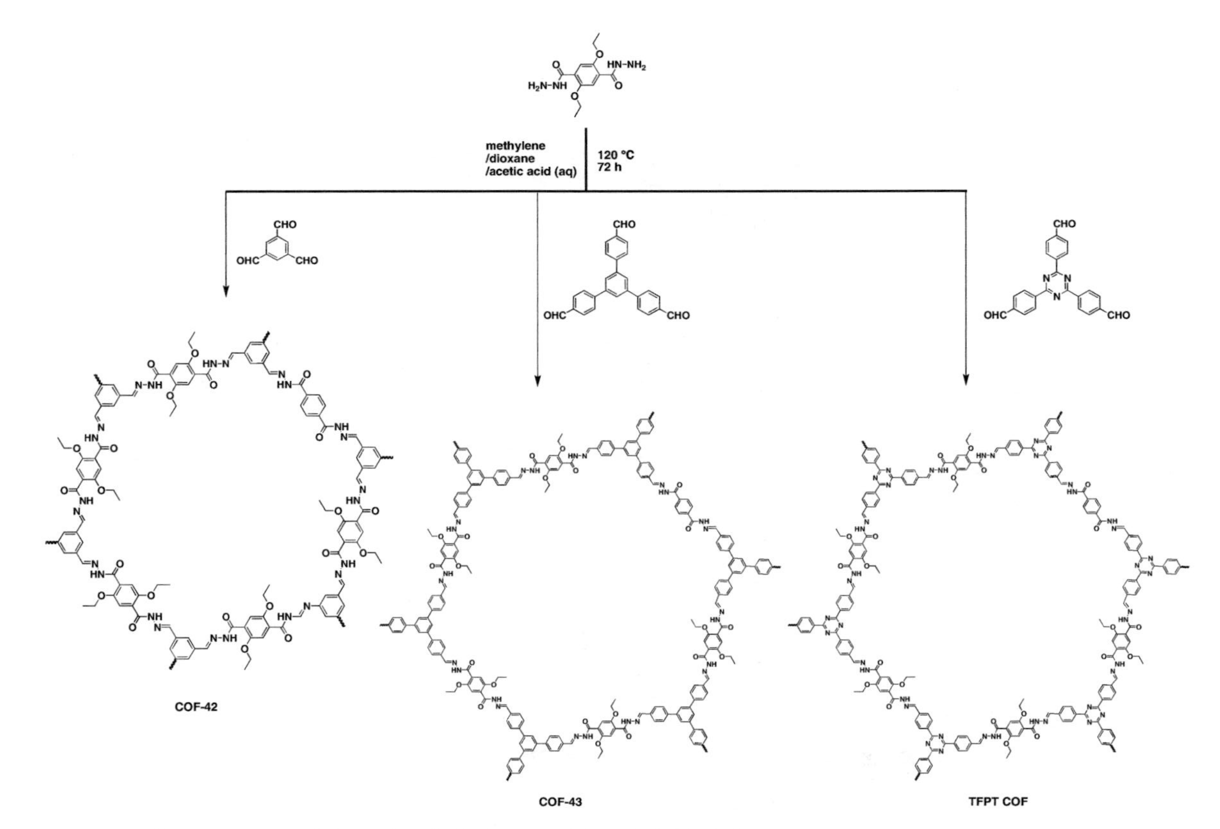

Figure 2.19 Synthesis of hydrazone-linked COFs (COF-42, COF-43, and TFPT COF).

2.4 Azine Linkages

$$2\ R_1CHO + H_2N\text{-}NH_2\ (\text{hydrazine}) \rightleftharpoons R_1CH{=}N\text{-}N{=}CHR_1 + 2\ H_2O \quad (2.7)$$

Azines (–C=N–N=C–) are formed by the condensation of hydrazine with aldehydes, as shown in Eq. 2.7. Azines are 2,3-diazosubstituted analogues of 1,3-dienes. One therefore might expect that they would show a noticeable π-electron conjugation between C=N bonds via the N–N linkage.

Usually, azine-linked COFs have small micropores and are stable in water, acid, and base (Fig. 2.20). Nagai et al. constructed a highly crystalline 2D azine-linked Py COF (Py-azine COF) by the condensation of hydrazine with 1,3,6,8-tetrakis(4-formylphenyl) pyrene under solvothermal conditions [67]. The azines inside of a Py-azine COF serve as Lewis basic sites to bind the guest and enable the selective detection of 2,4,6-trinitrophenol, which exhibits a potential application in chemosensing systems. Li et al. prepared an azine-linked COF, ACOF-1, by the condensation of hydrazine hydrate and 1,3,5-triformylbenzene. The high surface area (1318 m^2/g) and small pore size make it useful as a gas storage medium for CO_2 (177 mg/g at 273 K and 1 bar), H_2 (9.9 mg/g at 273 K and 1 bar), and CH_4 (11.5 mg/g at 273 K and 1 bar) [68, 69]. Smaldone and coworkers developed a novel azine-linked HPB-based COF, HEX-COF-1, which shows excellent sorption capability for CO_2 (20 wt%) and methane (2.3 wt%) at 273 K and 1 atm [70]. Lotsch and coworkers reported a tunable water- and photostable azine COF by hydrazine and triphenylarene aldehydes for visible light–induced hydrogen generation [71]. Zhang et al. reported an azo-containing COF by introducing a novel azobenzene monomer via the borate ester formation reaction of azobenzene-4,4′-diboronic acid and HHTP. The azo COF has a hexagonal skeleton and permanent porosity. The azo unit in azo COFs endow the COF material with photoisomerization properties [72].

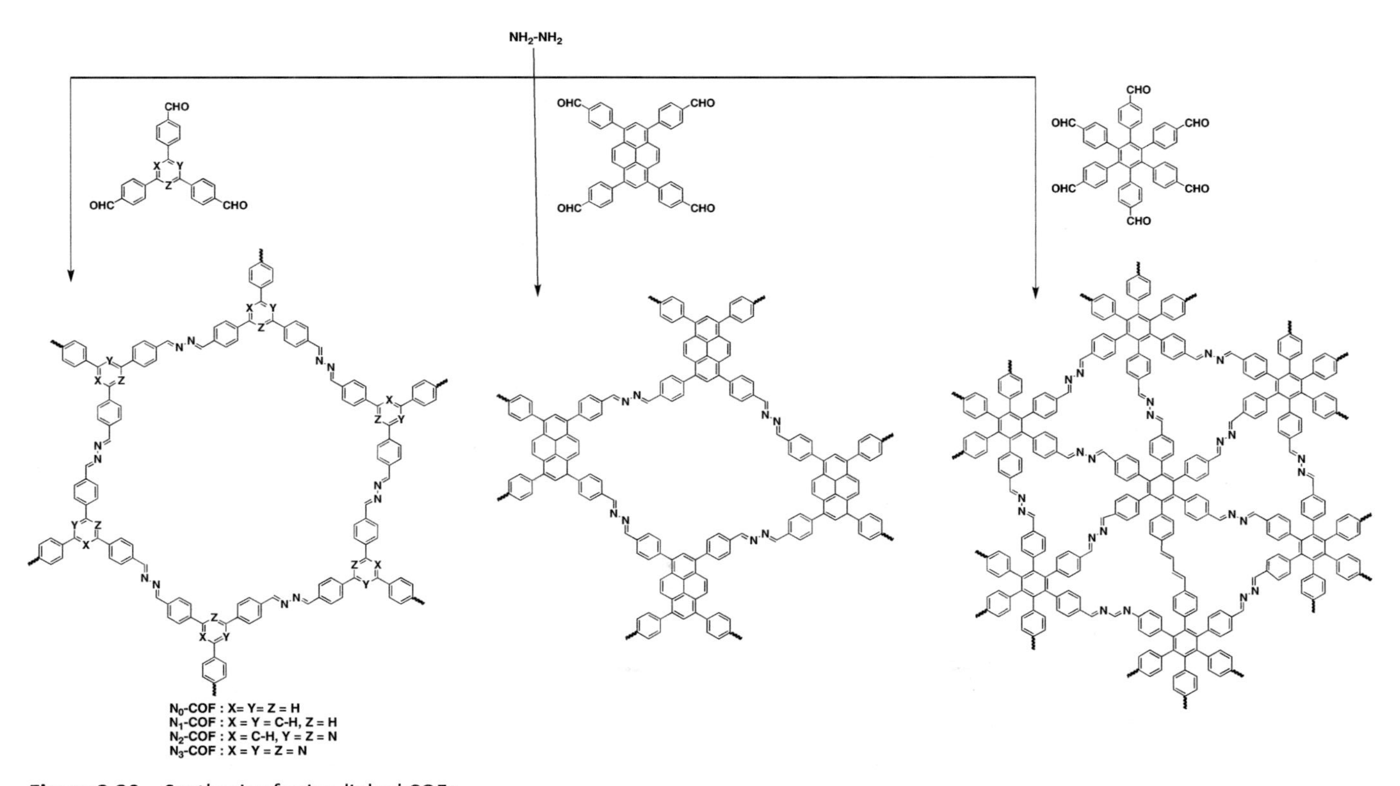

Figure 2.20 Synthesis of azine-linked COFs.

2.5 Squaraine Linkages

squaric acid $\xrightleftharpoons{2\ H_2N\text{-}R}$ squaraine $+ 2\ H_2O$ (2.8)

Squaric acid and amines are used as building blocks for the synthesis of squaraine-based COFs, which allow strong π conjugation from donor and acceptor interaction into a squaraine bond (Eq. 2.8). Nagai et al. reported a new reaction based on squaraine for the synthesis of a new type of building block to construct a crystalline 2D conjugated COF (CuP-Sq COF) with a tetragonal mesoporous skeleton (Fig. 2.21) [73]. The CuP-Sq COF possesses strong hydrogen bond capabilities that are strengthened by the delocalization of the lone pair of electrons in nitrogen into a four-membered ring to form zwitterionic resonance structures and shows strong π conjugation, as shown in Eq. 2.8. This COF shows a BET surface area of 539 m^2/g with a pore size of 2.1 nm. This research expands the types of COFs.

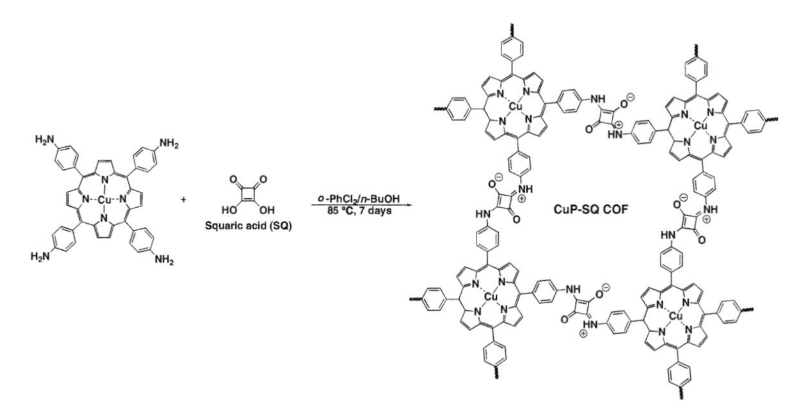

Figure 2.21 Construction of squaraine-based COFs (CuP-SQ COF).

2.6 Imide Linkages

$H_2N\text{-}R$ + anhydride $\rightleftharpoons$ imide $+ H_2O$ (2.9)

In organic chemistry, an imide bond is a functional group consisting of two acyl groups bound to nitrogen and is prepared by the condensation reaction of anhydrides and primary amines (Eq. 2.9). This imide-bonded compound is structurally related to acid anhydride. However, imide bonds are thermally reversible and more resistant toward hydrolysis.

Polyimides are formed by condensation of primary amines and anhydrides (Fig. 2.22). Tan and coworkers prepared 3D porous crystalline polyimide COFs (PI COFs) by choosing tetrahedral building blocks of different sizes [74]. These PI COFs show high thermal stability (more than 450°C) and large surface areas (more than 1000 m^2/g), as well as narrow pore sizes (13 Å for PI-COF-4 and 10 Å for PI-COF-5). They were the first COFs to be employed in controlled drug delivery. The Tan group also prepared a series of polyimide-based COFs with different pore sizes by changing the length of triamines using the imidization reaction [75]. Among them, PI-COF-3, synthesized by the condensation of pyromellitic dianhydride and the extended triamine, 1,3,5-tris[4-amino(1,1-biphenyl-4-yl)]benzene, has 5.3 nm wide pores and was loaded with rhodamine B, a dye probe for biological applications.

2.7 Phenazine Linkages

$$\text{C}_6\text{H}_4(\text{NH}_2)_2 + \text{C}_6\text{H}_4\text{O}_2 \rightleftharpoons (\text{C}_6\text{H}_4)_2\text{N}_2 + 2\,\text{H}_2\text{O} \qquad (2.10)$$

Phenazines as one of azo-fused structures were pursued in making conductive materials with the formula $(C_6H_4)_2N_2$ that have many potential applications in electrochemical devices [76]. The DCC reaction is led by the oxidation of dihydrophenazine, which is prepared by heading pyrocatechin with *o*-phenylenediamine (Eq. 2.10). However, the great challenge is how to make them highly crystalline. Jiang and coworkers prepared a phenazine-linked COF (CS COF, Fig. 2.23) via a ring-fusing reaction between quinone and amine derivatives catalyzed by acidic forms. They used C_3-symmetric triphenylene hexamine and C_2-symmetric *tert*-butylpyrene tetraone as building blocks to prepare the crystalline phenazine-linked CS COF under solvothermal conditions. The polygon skeletons of the COF are highly π conjugated and chemically stable in various solvents [77].

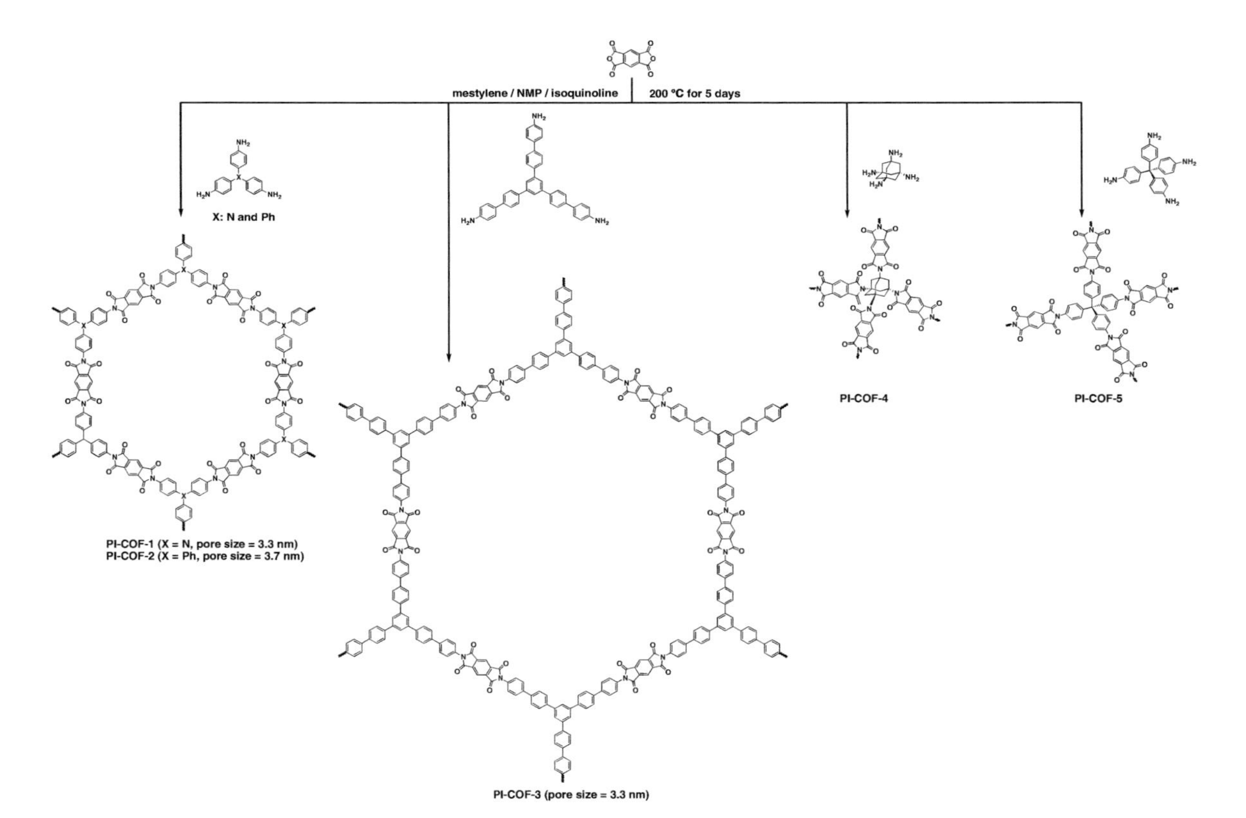

Figure 2.22 Development of 2D or 3D imide-linked COFs (PI-COF-1, PI-COF-2, PI-COF-3, PI-COF-4, and PI-COF-5) by the condensation of benzene-1,2,4,5-tetracarboxylic dianhydriodes.

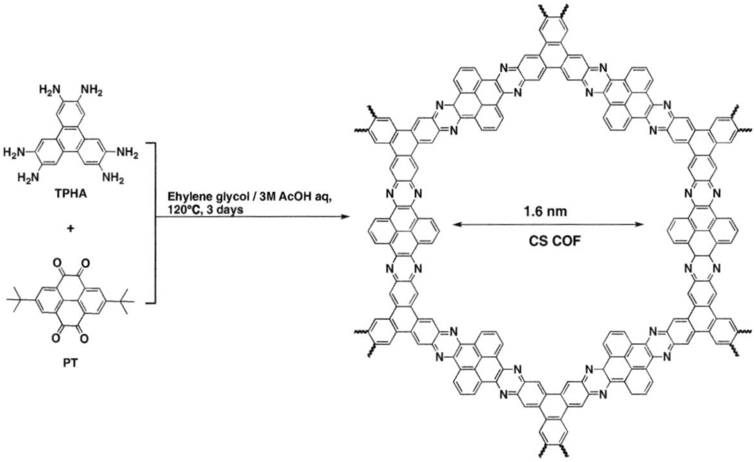

Figure 2.23 Synthesis of a phenazine-linked COF.

2.8 Triazine Linkages

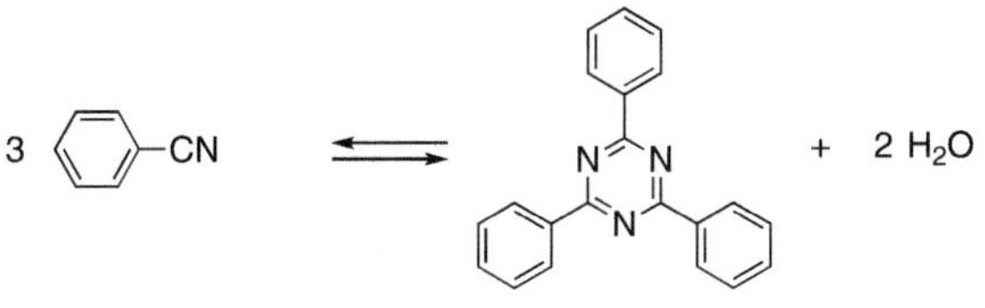

(2.11)

Triazine as 1,3,5-trizaine is a class of nitrogen-containing heterocycles and is synthesized by the condensation of aromatic nitriles at high temperatures in molten salts or at RT in the presence of a strong acid catalyst (Eq. 2.11). The crystalline, porous covalent triazine frameworks (CTFs) are prepared by condensational growth of aromatic nitriles in the presence of trifluoromethylsulfonic acid or zinc chloride, as shown in Fig. 2.24.

The CTF was initially reported by Thomas and coworkers in 2008 and is prepared via the dynamic trimerization of aromatic nitriles (1,3,5-triazines) using the $ZnCl_2$ ionothermal method [78]. CTFs gave high surface areas (up to 2475 m^2/g), low densities, and outstanding thermal/chemical stability. However, CTFs show short-range crystalline order and limited pore size distributions in some cases [79].

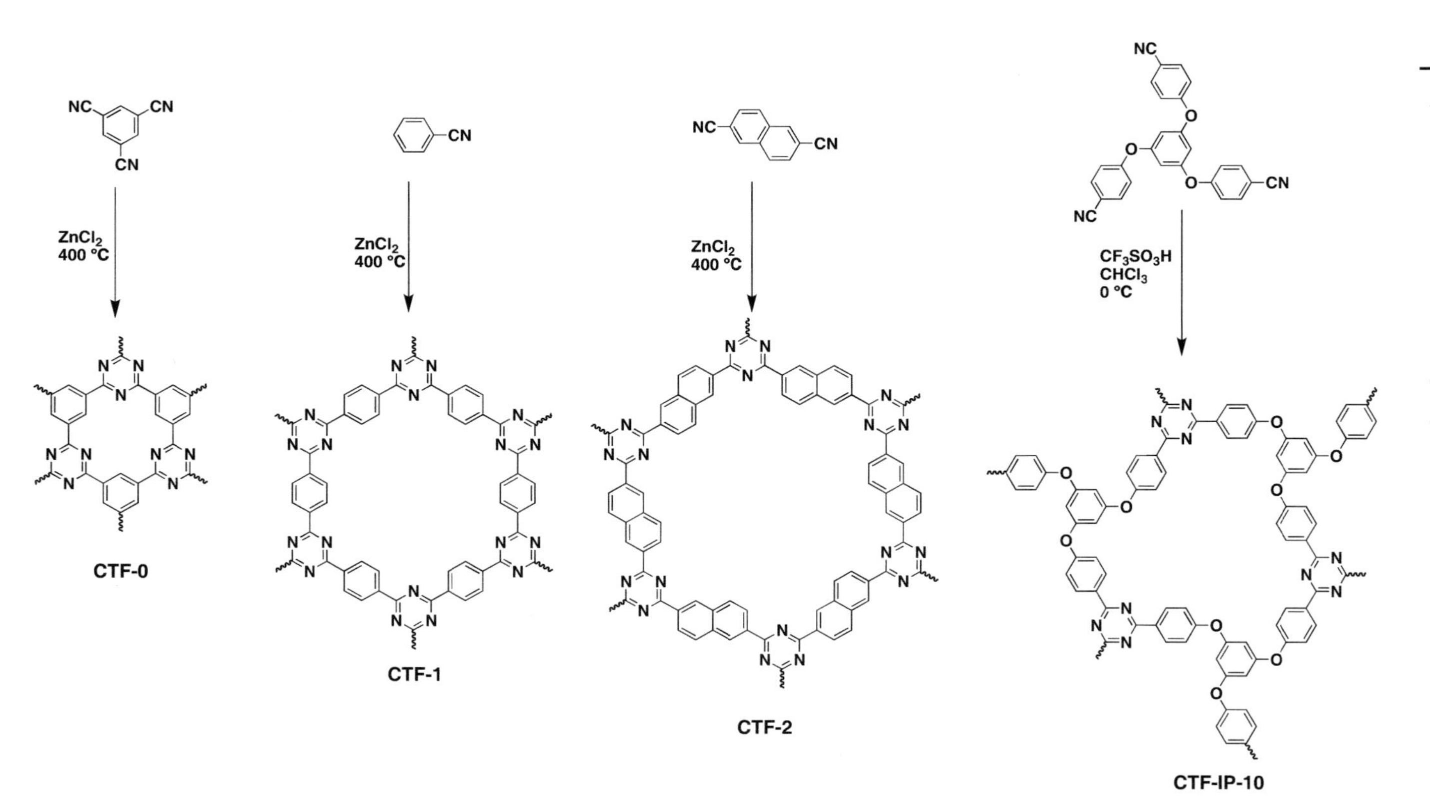

Figure 2.24 Construction of crystalline covalent triazine frameworks (CTFs) by self-condensation of aromatic nitriles.

Most CTFs were synthesized in molten $ZnCl_2$ at high temperatures (400°C–700°C). These conditions for CTF synthesis provided less crystalline CTFs of limited functionality and only from 1,4-dicyanobenzene and 2,6-dicyanobenzene [80]. Then, Cooper et al. reported a microwave-assisted, acid-catalyzed method to synthesize triazine networks with fewer discolored impurities (P1M-P6M). Their structural and synthesized tunability developed their specific applications, such gas storage, catalyst support, and organic dye separation [81].

Thomas at al. reported the synthesis of organic CTFs (CTF-0) by the trimerization of 1,3,5-trisyanobenzene in molten $ZnCl_2$ [82]. This COF can present high CO_2 uptakes and good catalytic activity of CO_2 cycloaddition. The high BET surface area and robust porous structures of the organic framework were explored for highly selective gas capture.

Ghosh and coworkers found a strategy to obtain a porous CTF (CTF-IP-10) with bimodal functionality by acid-catalyzed RT reaction of tricyanomonomer (PCN-M1) in $CHCl_3$ [83]. The obtained CTF-IP-10 consists of both an electron-deficient central triazine core and electron-rich aromatic building blocks.

2.9 Multihetero Linkages in One COF Skeleton

To improve the functionality and stability of the COF structure with b–O linkages, COFs connected by two or more types of covalent bonds on orthogonal reaction were designed and synthesized.

First Zhao and group reported a 2D COF (NTU-COF-1, Fig. 2.25) by the formation of imine group and boroxine (B_3O_3) rings, involving the use of two building blocks: 4-formyl-phenylboronic acid (FPBA) and 1,3,5-tris(4-aminophenyl)benzene. They also constructed a ternary COF (NTU-COF-2, Fig. 2.25) via the formation of two types of covalent bonds. The NTU-COF-2 has a high BET surface area (1619 m^2/g) and mesopores with a diameter of 2.6 nm. The design and construction of COFs by this orthogonal reaction strategy is a promising approach in rational COF development [84].

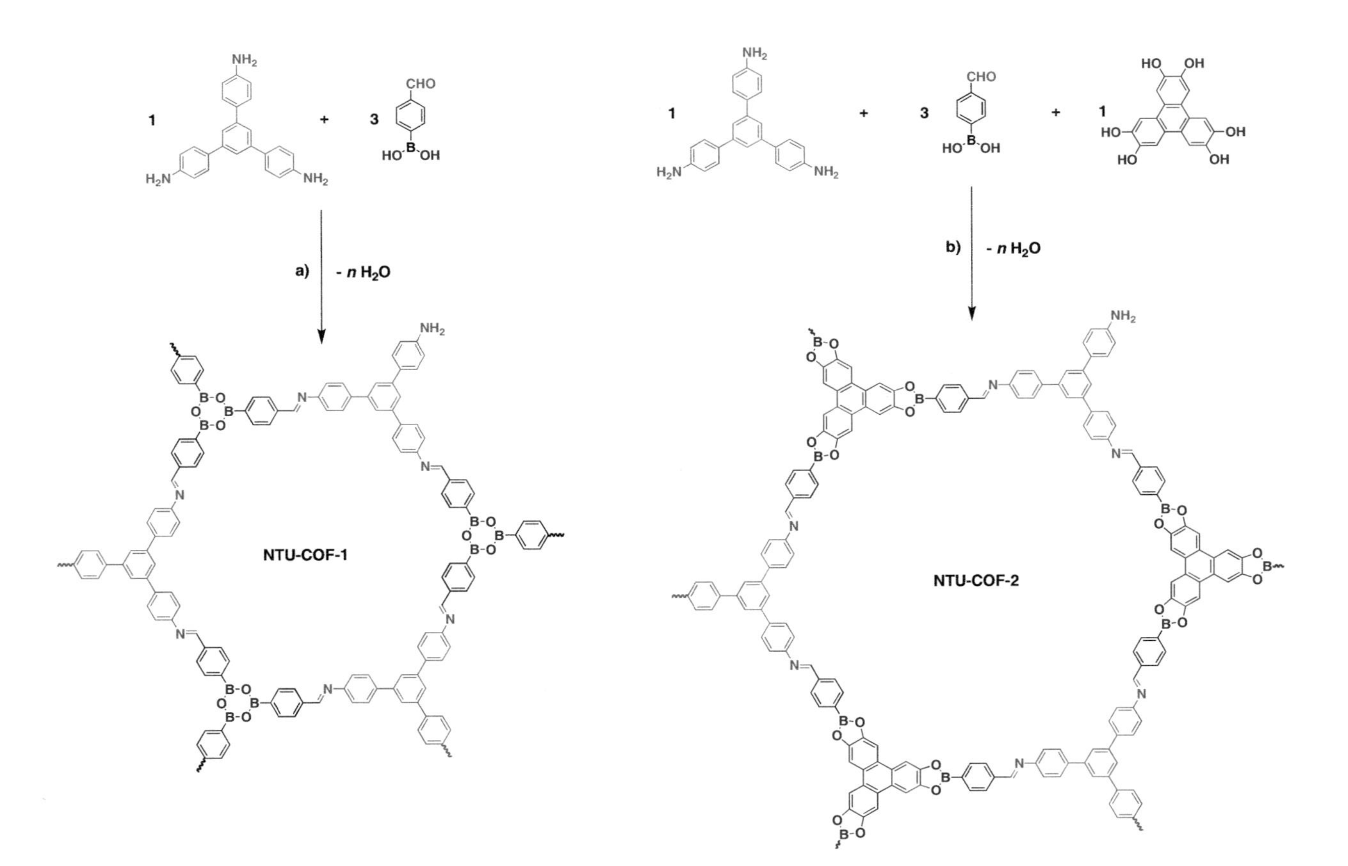

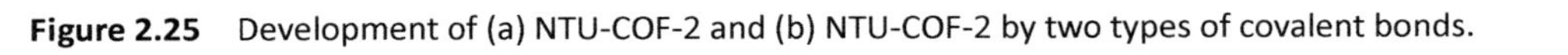

Figure 2.25 Development of (a) NTU-COF-2 and (b) NTU-COF-2 by two types of covalent bonds.

Qiu's group prepared two extended 3D COFs (DL-COF-1 and DL-COF-2) with dual linkages, imine groups and B_3O_3 boroxine rings (Fig. 2.26), involving the utilization of two 1,3,5,7-tetraaminoadamantane and FPBA or 2-fluoro-4-formylphenyl boronic acid building blocks [85]. The obtained DL-COF-1 and DL-COF-2 showed high BET surface areas (2259 m^2/g and 2071 m^2/g, respectively).

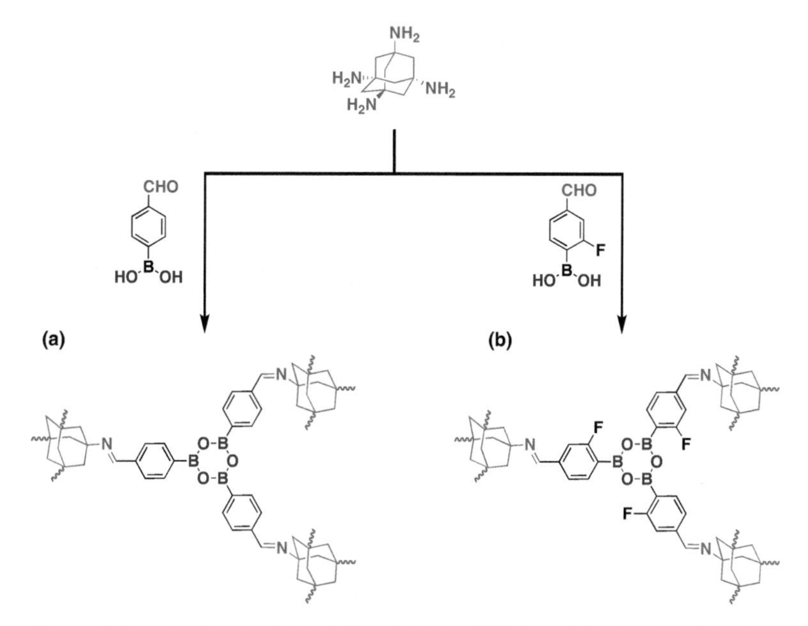

Figure 2.26 Synthesis of 3D COFs of (a) DL-COF-1 and (b) DL-COF-2 by two types of covalent bonds from orthogonal reactions.

2.10 Perspectives and Challenges

The past decade has witnessed a boom in COF chemistry. Their unique features, such as flexible molecular structure design, permanent porosity, low density, modest to high crystallinity, and controllable pore size (from ultraporous to mesoporous) and the diversity of the available building blocks, easily make them targets for structural and functional design and promise their great potential as materials for application in various areas. However, there are still a lot of challenges in the design and synthesis of COFs.

First, compared to the highly ordered structure of MOF materials, further investigation on thermodynamically controlled solvothermal reactions for the preparation of COF materials is quite important. Such studies provide a general pathway for the preparation of high-crystallinity COFs. Moreover, control over the defect sites, morphology, and stacking mode of COF structures is also important and effective methods to construct precise crystal COF structures are necessary. At present, no direct analytical tools can be used to visualize and analyze the stacking and topological structures of COFs. Thus, the creation of single-crystal COFs is an unattainable target.

Meanwhile, in the existing approaches, the methods to increase the surface areas and porosity for COF materials are a big bottleneck. The surface areas and porosity are the most important parameters for COF materials for gas adsorption and gas separation applications. How to design COF structures with improved surface area and these properties is the key to realizing the potential application of COFs in these fields. For 2D COFs, spatially enlarging the interlayered spaces would increase the surface areas. For 3D COFs, the construction of highly crystalline frameworks would lead to very high BET surface areas, which can reach the theoretical S_{BET} of over 6000 $\mathrm{m^2/g}$.

References

1. Korich, A. L., Inovine, P. M. (2010). *Dalton Trans.*, **39**, 1423–1431.
2. Scheleyer, P. V., Jiao, H. J., Hommes, N. J. R. V., Malkin, V. G., Malkina, O. L. (1997). *J. Am. Chem. Soc.*, **119**, 12669–1779.
3. Brock, C. P., Minton, R. P., Niedenzu, K. (1987). *Acta Crystallogr.*, **C43**, 1257–1268.
4. Synder, H. R., Kuck, J. A., Johnson, J. R. (1938). *J. Am. Chem. Soc.*, **60**, 105–111.
5. Morgan, A. B., Jurs, J. L., Tour, J. M. (2000). *J. Appl. Polym. Sci.*, **76**, 1257–1268.
6. Li, Y., Ding, J., Day, M., Tao, T., Lu, J., D'iorio, M. (2003). *Chem. Mater.*, **15**, 4936–4934.
7. Tokunaga, Y., Ueno, H., Shimomura, Y., Seo, T. (2002). *Heterocycles*, **57**, 787–790.

8. Côté, A. P., Benin, A. I., Ockwig, N. W., O'Keefe, M., Matzager, A. J., Yaghi, O. M. (2005). *Science*, **310**, 1166–1170.
9. El-Kaderi, H. M., Hunt, J. R., Mendoza-Cortez, A. P., Côté, A. P., Taylor, R. M., O'Keefe, M., Yaghi, O. M. (2007). *Science*, **316**, 268–272.
10. Wan, S., Guo, J., Kim, J., Ihee, H., Jiang, D. (2009). *Angew. Chem. Int. Ed.*, **48**, 5439–5442.
11. Campbell, N. L., Clowes, R., Ritchie, L. K., Cooper, A. I. (2009). *Chem. Mater.*, **21**, 204–206.
12. Côté, A. P., Benin, A. I., Ockwig, N. W., O'Keefe, M., Matzager, A. J., Yaghi, O. M. (2005). *Science*, **310**, 1166–1170.
13. El-Kaderi, H. M., Hunt, J. R., Mendoza-Cortez, A. P., Côté, A. P., Taylor, R. M., O'Keefe, M., Yaghi, O. M. (2007). *Science*, **316**, 268–272.
14. Côté, A. P., El-Kaderi, H. M., Furukawa, H., Hunt, J. R., Yaghi, O. M. (2007). *J. Am. Chem. Soc.*, **129**, 12914–12915.
15. Wan, S., Guo, J., Kim, J., Ihee, H., Jiang, D. (2009). *Angew. Chem. Int. Ed.*, **48**, 5439–5442.
16. Tilford, R. W., Mugavero, S. J., Pellechia, P. J., Lavigne, J. J. (2008). *Adv. Mater.*, **20**, 2741–2746.
17. Lanni, L. M., Tilford, R. W., Bharathy, M., Lavigne, J. J. (2007). *J. Am. Chem. Soc.*, **133**, 13975–13983.
18. Dogru, M., Sonnauer, A., Gavryushin, A., Knochel, P., Bein, T. A. (2011). *Chem. Commun.*, **133**, 13975–13983.
19. Spitler, E. L., Colson, J. W., Uribe-Romo, F. J., Woll, A. R., Giovino, M. R., Saldivar, A., Ditchel, W. R. (2012). *Angew. Chem. Int. Ed.*, **51**, 2623–2627.
20. Colson, J. W., Mann, J. A., DeBlase, C. R., Dichtel, W. R. (2015). *J. Polym. Sci. Polym. Chem.*, **53**, 378–384.
21. Spitler, E. L., Dichtel, W. R. (2010). *Nat. Chem.*, **2**, 672–677.
22. Spitler, E. L., Giovino, M. R., White, S. L., Dichtel, W. R. (2011). *Chem. Sci.*, **2**, 1588–1593.
23. Dogru, M., Sonnauer, A., Zimdars, S., Doblinger, M., Knochel, P., Bein, T. (2013). *Crystengcomm*, **15**, 1500–1502.
24. Calik, M., Sick, T., Dogru, M., Döblinger, M., Datz, S., VUdde, H., HArtschuh, A., Auras, F., Bein, T. (2016). *J. Am. Chem. Soc.*, **138**, 1234–1239.
25. Böeseken, J. (1949). *Adv. Carbohydrate Chem.*, **4**, 189–210.
26. Hermans, P. H. (1925). *Z. Anorg. Allgem. Chem.*, **142**, 83–110.
27. (a) Xiao, L., Ling, Y., Alsbaiee, A., Li, D., Helbling, D. E., Dichtel, W. R. (2017). *J. Am. Chem. Soc.*, **139**, 7689–7692; (b) Alsbaiee, A., Smith, B.

J., Ling, Y., Li, Y., Helbling, D. E., Dichtel, W. R. (2016). *Nature*, **529**, 190–194.

28. Smsldone, R. A., Forgan, R. S., Furukawa, H., Gassensmith, J. J., Slawin, A. M., Yaghi, O. M., Stoddart, J. F. (2010). *Angew. Chem. Int. Ed.*, **49**, 8630–8634.
29. Brown, H. C., Rangaishevi, M. V. (1988). *J. Organimet. Chem.*, **358**, 15–30.
30. Van Duin, M., Peters, J. A., Kleboom, A. P. G., Van Bekkum, H. (1985). *Tetrahedron*, **41**, 3411–3421.
31. Van Duin, M., Peters, J. A., Kleboom, A. P. G., Van Bekkum, H. (1984). *Tetrahedron*, **40**, 2901–2911.
32. Yoshino, K., Kotaka, M., Okamoto, M., Kakihara, H. (1979). *Bull. Chem. Soc. Jpn.*, **52**, 3005–3009.
33. Voisin, E., Maris, T., Wuest, J. D. (2008). *Cryst. Growth Des.*, **8**, 308–318.
34. Danjo, H., Hirata, K., Yoshiga, S., Azumaya, I., Yamaguchi, K. (2009). *J. Am. Chem. Soc.*, **131**, 1638–1639.
35. Loewer, Y., Weiss, C., Biju, A. T., Froehilich, R., Glorius, F. (2011). *J. Org. Chem.*, **76**, 2324–2327.
36. Abrahams, B. F., Price, D. J., Robson, R. (2006). *Angew. Chem. Int. Ed.*, **45**, 806–810.
37. Zhang, Y., Duan, J., Ma, D., Li, P., Li, S., Li, H., Zhou, J., Ma, Z., Feng, Z., Wang, B. (2017). *Angew. Chem. Int. Ed.*, **65**, 16313–16317.
38. (a) Kim, D. P., Economy, L. G. (1994). *Chem. Mater.*, **6**, 395–400; (b) Fazen, P. L., Remsen, E. E., Beck, J. S., Carroll, P. J., McGhie, A. R., Sneddon, L. G. (1995). *Chem. Mater.*, **7**, 1942–1956.
39. (a) Lynch, A. T., Sneddon, J. G. (1989). *J. Am. Chem. Soc.*, **111**, 6201–6209; (b) Riedel, R., Bill, J., Passing, G. (1991). *Adv. Mater.*, **3**, 551–552; (c) Bonnetot, B., Guilhon, F., Viala, J. C., Mongeot, H. (1995). *Chem. Mater.*, **7**, 299–303.
40. (a) Schaeffer, G. W., Anderson, E. R. (1949). *J. Am. Chem.*, **71**, 2143–2145; (b) Hougn, W. V., Schaeffer, G. W., Dzurns, M., Stewart, A. C. J. (1955). *J. Am. Chem.*, **77**, 864–865
41. Jaska, C. A., Temple, K., Lough, A. J., Manners, I. (2003). *J. Am. Chem.*, **125**, 9424–9434.
42. Framery, E., Vaultier, M. (2000). *Heteroat. Chem.*, **11**, 218–225.
43. Wakamiya, A., Ide, T., Yamaguchi, S. (2005). *J. Am. Chem.*, **127**, 14859–14866.

44. Jackson, K. T., Reich, T. E., El-kaderi, H. M. (2012). *Chem. Commun.*, **48**, 8823–8825.

45. Uribe-Romo, F. J., Hunt, J. R., Furukawa, H., Klock, C., O'Keefe, M., Yaghi, O. M. (2009). *J. Am. Chem. Soc.*, **131**, 4570–4571.

46. Zhang, Y. B., Su, J., Furukawa, H., Yun, Y. F., Gandara, F., Duong, A., Zou, X. D., Yaghi, O. M. (2013). *J. Am. Chem. Soc.*, **135**, 16336–16339.

47. Ma, Y. X., Li, Z. J., Wei, L., Ding, S. Y., Zhang, Y. B., Wang, W. A. (2017). *J. Am. Chem. Soc.*, **139**, 4995–4998.

48. Lin, G. Q., Ding, H. M., Yuan, D. Q., Wang, B. S. (2016). *J. Am. Chem. Soc.*, **138**, 3302–3305.

49. Fang, Q. R., Gu, S., Zheng, J., Zhuang, Z. B., Qiu, S. L., Yan, Y. S. (2014). *Angew. Chem. Int. Ed.*, **53**, 2878–2882.

50. Kandambeth, S., Shinde, D. B., Panda, M. K., Lukose, B., Heine, T., Banerjee, R. (2013). *Angew. Chem. Int. Ed.*, **52**, 13052–13056.

51. Chen, X., Addicoat, M., Jin, E., Zhai, L., Xu, H., Huang, N., Guo, Z., Liu, L., Irle, S., Jiang, D. (2015). *J. Am. Chem. Soc.*, **137**, 3241–3247.

52. Chen, X., Addicoat, M., Irle, S., Nagai, A., Jiang, D. (2013). *J. Am. Chem. Soc.*, **135**, 546–549.

53. Xu, H., Gao, J., Jiang, D. (2015). *Nat. Chem.*, **7**, 905–912.

54. Ding, X., Chen, Y., Honsho, Y., Feng, X., Saengawang, O., Guo, J., Saeki, A., Seki, S., Irle, S., Nagase, S., Parasuk, V., Jiang, D. (2011). *J. Am. Chem. Soc.*, **133**, 14510–14513.

55. Song, J. R., Sun, J., Liu, J., Huang, Z. T., Zheng, Q. Y. (2014). *Chem. Commun.*, **50**, 788–791.

56. Lammi, L. M., Tilford, R. W., Bharathy, M., Lavigne, J. J. (2011). *J. Am. Chem. Soc.*, **133**, 13975–13983.

57. Spitler, E. L., Koo, B. T., Novotney, J. L., Colson, J. W., Uribe-Romo, F. J., Gutierrez, G. D., Clancy, P., Dichtel, W. R. (2011). *J. Am. Chem. Soc.*, **133**, 19416–19421.

58. Dalapati, S., Addicoat, M., Jin, S., Sakurai, T., Gao, J., Xu, H., Irle, S., Seki, S., Jiang, D. (2015). *Nat. Commnun.*, **6**, 7786.

59. Kandambeth, S., Mallick, A., Lukose, B., Mane, M. V., Heine, T., Banerjee, R. (2012). *J. Am. Chem. Soc.*, **134**, 19524–19527.

60. Biswal, B. P., Chandra, S., Kandambeth, S., Lukose, B., Heine, T., Banerjee, R. (2013). *J. Am. Chem. Soc.*, **135**, 5328–5331.

61. Waller, P. J., Lyle, S. J., Popp, T. M. O., Diercks, C. S., Reimer, J. A., Yaghi, O. M. (2016). *J. Am. Chem. Soc.*, **138**, 15519–15522.

62. (a) Job, A., Janeck, C., Bettary, W., Peters, R., Enders, D. (2002). *Tetrahedron*, **58**, 2253–2329; (b) Sugiura, M., Kobayashi, S. (2005). *Angew. Chem. Int. Ed.*, **44**, 5176–5186.
63. Uribe-Romo, F. J., Doonan, C. J., Furukawa, H., Oisaki, K., Yaghi, O. M. (2011). *J. Am. Chem. Soc.*, **1338**, 11478–11481.
64. Bunk, D. N., Dichtel, W. R. (2013). *J. Am. Chem. Soc.*, **135**, 14952–14955.
65. Stegbauer, L., Schwinghammer, K., Lotsch, B. V. (2014). *Chem. Sci.*, **5**, 2789–2793.
66. Ding, S. Y., Dong, M., Wang, Y. W., Chen, Y. T., Wang, H. Z., Su, C. Y., Wang, W. (2016). *J. Am. Chem. Soc.*, **138**, 3031–3037.
67. Dalapati, S., Jin, S. B., Gao, Y. T., Xu, T. C., Nagai, A., Jiang, D. L. (2013). *J. Am. Chem. Soc.*, **135**, 17310–17313.
68. Li, Z. P., Feng, X., Zou, Y. C., Zhang, Y. W., Xia, H., Mu, Y. A. (2014). *Chem. Commun.*, **50**, 13825–13828.
69. Li, Z. P., Zhi, Y. F., Feng, X., Ding, X. S., Zou, Y. C., Liu, X. M., Mu, Y. (2015). *Chem.-Eur. J.*, **21**, 12079–12084.
70. Alahakoon, S. B., Thompson, C. M., Nguyen, A. X., Occhialini, G., McCandless, G. T., Smaldone, R. A. (2016). *Chem. Commun.*, **52**, 2843–2845.
71. Vyas, V. S., Haase, F., Stegbauer, L., Savasci, G., Podjaski, F., Ochsenfeld, C., Lotsch, B. V. (2015). *Nat. Commun.*, **6**, 8508–8516.
72. Zhang, J., Wang, L. B., Li, N., Liu, J. F., Zhang, Z. B., Zhou, N. C., Zhu, X. L. (2014). *Crystengcomm*, **16**, 6547–6551.
73. Nagai, A., Chen, X., Feng, X., Ding, X., Guo, Z., Jiang, D. (2013). *Angew. Chem. Int. Ed.*, **52**, 3770–3774.
74. Fang, Q. R., Wang, J. H., Gu, S., Kaspar, R. B., Zhang, J., Wang, J. H., Qiu, S. L., Tan, Y. S. (2013). *J. Am. Chem. Soc.*, **137**, 8352–8355.
75. Fang, Q. R., Zhuang, Z. B., Gu, S., Kaspar, R. B., Zhang, J., Wang, J. H., Zhang, J., Wang, J. H., Qiu, S. L., Tan, Y. S. (2014). *Nat. Commun.*, **5**, 4503–4510.
76. Kou, Y., Xu, Y. H., Guo, Z. Q., Jiang, D, L. (2011). *Angew. Chem. Int. Ed.*, **50**, 8753–8757.
77. Guo, J., Xu, Y., Jin, S., Chen, L., Kaji, T,; Honsho, Y., Addicoat, M. A., Kim, J., Saeki, A., Ihee, H., Seki, S., Irle, S., Hiramoto, M., Gao, J., Jiang, D. (2013). *Nat. Chmmun.*, **4**, 2736–2743.
78. Kuhn, P., Antoietti, M., Thomas, A. (2008). *Angew. Chem. Int. Ed.*, **47**, 3450–3453.

79. Kuhn, P., Forget, A., Su, D., Thomas, A., Antonietti, M. (2008). *J. Am. Chem. Soc.*, **130**, 13333–13337.

80. Bojdys, M. J., Jermenok, J., Thomas, A., Antonietti, M. (2010). *Adv. Mater.*, **22**, 2202–2205.

81. Ren, S. J., Bojdys, M. J., Dawson, R., Layboum, A., Khimyak, Y. Z., Adams, D. J., Cooper, A. I. (2012). *Adv. Mater.*, **24**, 2357–2361.

82. Katekomol, P., Roeser, J., Bojdys, M., Weber, J., Thomas, A. (2013). *Chem. Mater.*, **25**, 1542–1548.

83. Karmakar, A., Kumar, A., Chaudhari, A. K., Samanta, P., Desai, A. V., Krishna, R., Ghosh, S. K. (2016). *Chem.-Eur. J.*, **22**, 4931–4937.

84. Zeng, Y., Zou, R., Luo, Z., Zhang, H., Yao, X., Ma, X., Zou, R., Zhao, Y. (2015). *J. Am. Chem. Soc.*, **137**, 1020–1023.

85. Li, H., Pan, Q., Ma, Y., Guan, X., Xue, M., Fang, Q., Yan, Y., Valtchev, V., Oiu, S. (2016). *J. Am. Chem. Soc.*, **138**, 14783–14788.

Chapter 3

Gas Adsorption and Storage of COFs

COFs have been touted as a new class of porous crystalline materials for gas storage applications due to their high inherent surface areas, very low densities (being composed of light elements, such as H, C, N, B, O, and Si), high stability from the formation of covalent bonds, and tunable pore dimensions, ranging from ultramicropores to mesopores. The gas storage capacity of COFs for hydrogen, methane, and carbon dioxide has attracted increasing interest in the past decades [1]. As a general rule, the gas adsorption capacity of a COF depends primarily on the components and topologies of its frameworks. Because of their larger surface areas and pore volumes, 3D COFs possess significantly higher uptake capacities than 2D COFs. Moreover, COF-based thin film nanocomposite (TFN) membranes and freestanding COF films were introduced, with application in gas separation.

3.1 Gas Sorption

The true surface area, including surface irregularities and pore interiors, cannot be calculated from particle size information but is rather determined at the atomic level by the adsorption of an unreactive, or inert, gas. The amount adsorbed, let's call it X, is a function not only of the total amount of exposed surface but also of (i) temperature, (ii) gas pressure, and (iii) the strength of interaction

Covalent Organic Frameworks
Atsushi Nagai

ISBN 978-981-4800-87-7 (Hardcover), 978-1-003-00469-1 (eBook)
www.jennystanford.com

between gas and solid. Because most gases and solids interact weakly, the surface must be cooled substantially in order to cause a measurable amount of adsorption—enough to cover the entire surface. As the gas pressure is increased, more is adsorbed on the surface (in a nonlinear way). However, adsorption of a cold gas does not stop when it has covered the surface in a complete layer one-molecule thick (let's call the theoretical monolayer amount of gas X_m). As the relative pressure is increased, excess gas is adsorbed to form "multilayers"; thus, gas adsorption as a function of pressure does not follow a simple relationship, and we must use an appropriate mathematical model for the surface area.

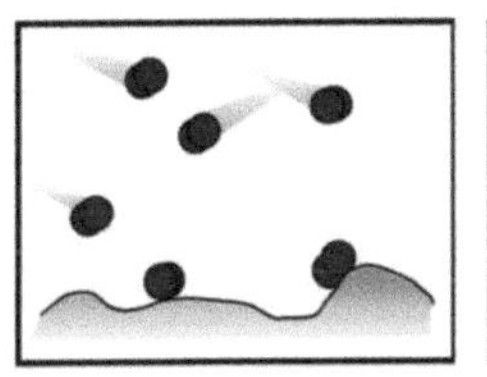
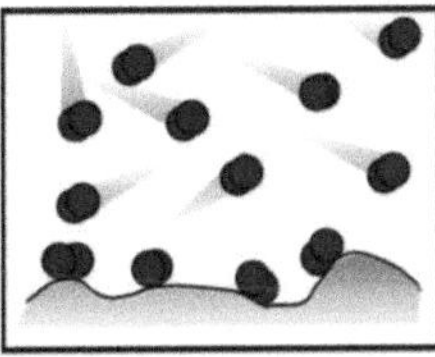
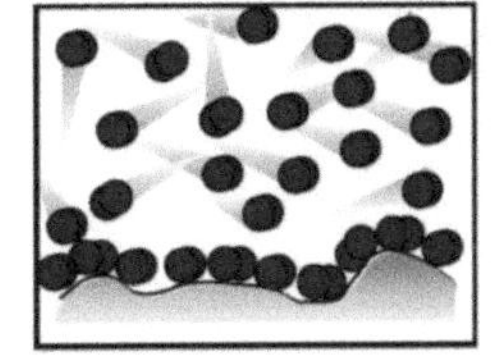

Figure 3.1 Schematic of gas adsorption.

3.2 Physical and Chemical Adsorption

Solid surfaces are not smooth in the microscopic sense owing to irregular valleys and peaks distributed all over the area. The regions of irregularity are particularly susceptible to residual force fields. At these locations, the surface atoms of the solid may attract other atoms or molecules in the surrounding gas or liquid phase. The surfaces of pure crystal have nonuniform force fields due to the atomic structure in the crystal. Such surfaces also have sites or active centers where adsorption is favored.

Adsorption on a solid surface may be divided into two categories, physical adsorption and chemical adsorption. Physical adsorption is nonspecific and similar to condensation. The forces that attract the fluid molecules to the solid surface are weak van der Waals forces. Therefore, physical adsorption is also known as van der Waals adsorption. The heat evolved during physical adsorption is low, which usually lies between 2 and 25 kJ/mol. The energy of activation for physical adsorption is also low (<5 kJ/mol). Equilibrium between the solid surface and the gas molecules is usually attained rapidly, and it is reversible because the energy requirements are small. Multiple layers of adsorbed molecules are possible, especially near the condensation temperature. The extent of physical adsorption decreases with increasing temperatures. Physical adsorption is useful for determining the surface area and pore size of a solid catalyst.

In contrast, chemical adsorption, or chemisorption, is specific. It involves forces that are much stronger than physical adsorption. The amount of heat evolved in chemisorption is large (e.g., 50–500 kJ/mol), which is similar to that in chemical adsorptions. The term "chemisorption" was given by Taylor [2]. However, the concept of chemisorption was proposed by Langmuir much earlier [3]. According to him, in the interior of the solid material (i.e., the adsorbent), the atoms have their force fields wholly satisfied by the atoms that surround them. The atoms on the surface, however, are not surrounded. These atoms have a residual force due to unshared electrons toward the exterior. This residual force field leads to the sharing of electrons with striking gas molecules (i.e., the adsorbate).

A sort of covalent linkage is formed between the gas molecules and the atoms of the solid at the surface. However, this linkage is not the same as the covalent bond that exists in a chemical compound. In spite of the sharing of the electron with the gas molecule, the surface atom remains bonded simultaneously with the other atoms in the crystal lattice.

Langmuir observed that a stable film of oxide formed on tungsten wires in the presence of oxygen, which was not the same as the normal oxide, WO_3, in terms of properties. This oxide was given off from the surface upon desorption. This adsorbed compound was designated as $W \bullet O_3$. Langmuir pointed out that because of the rapid falling off of intermolecular forces with distance, it is probable that the adsorbed layers are no more than a single molecule in thickness. Langmuir proposed simple formulations for rates of adsorption and desorption of gases on a solid surface [4]. These are also applicable to liquids. Unlike physical adsorption, chemisorption does not take place on all solids. It occurs with some chemically reactive gases. Chemisorption usually occurs at high temperatures. The surface coverage is limited to a monolayer. The process is often irreversible. Chemisorption is used for finding the active centers of a catalyst. Chemisorption can be divided into two categories, activated chemisorption and nonactivated chemisorption. In activated chemisorption, the rate of chemisorption varies with temperature following Arrhenius law with finite activation energy. In some situations, chemisorption occurs rapidly, which suggests that the activation energy is nearly zero. This type of adsorption is called nonactivated chemisorption. Both activated and nonactivated chemisorption may take place at different stages of an adsorption process. A comparison between physical adsorption and chemisorption is presented in Table 3.1.

Table 3.1 Comparison between physical adsorption and chemisorption

Physical adsorption	Chemisorption
All solid can be used as adsorbents.	Some solids can be used as adsorbents.
All gases below their critical temperatures can act as adsorbents.	Some chemically reactive gases can act as adsorbents.
It occurs at low temperatures.	It generally occurs at high temperatures.
The heat of adsorption is low.	The heat of reaction is high, which is on the order of the heat of chemical reactions.
The rate is high.	Both low and high rates are observed.
The activation energy is low.	The activation energy is low for nonactivated chemisorption and high for activated chemisorption.
A multilayer is possible.	Only a monolayer is formed.
The reaction is highly reversible.	The reaction is often irreversible.
It is used for the determination of surface area and pore size.	It is used for the determination of the active center area and elucidation of surface reaction kinetics.

3.3 Brunauer–Emmett–Teller Theory

The theory of chemical adsorption proposed by Langmuir is based on the formation of a monolayer. However, multilayers like covalent organic frameworks (COFs) can form in physical adsorption. For example, several layers of nitrogen molecules can get adsorbed on top of each other on the surface of silica gel at 77 K below atmosphere pressure. The isotherm in the case of a monolayer assumes the shape shown in Fig. 3.2 (type I).

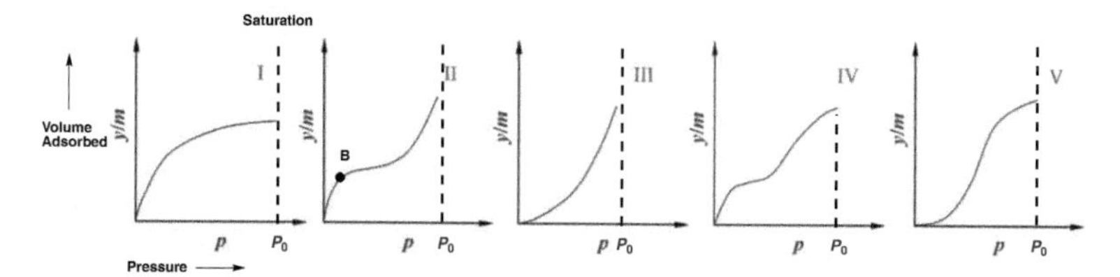

Figure 3.2 Adsorption isotherms.

The extent of adsorption increases with pressure and ultimately reaches a limiting value, as predicted by the Langmuir theory. However, for adsorption involving multilayer formation at low temperature, at least four other types of adsorption isotherms can be observed (types II–V in Fig. 3.1). Type II is perhaps the most common for physical adsorption on relatively open surfaces, in which adsorption proceeds progressively from submonolayer to multilayer; the isotherm exhibits a distinct concave downward curvature at some low relative pressure (P/P_0) and a sharply rising curve at high P/P_0. Point B at the knee on the curve signifies completion of an adsorbed monolayer. It forms the basis of the Brunauer–Emmett–Teller (BET) model for surface area determination of a solid from the assumed monolayer capacity, described below.

A type III isotherm signifies a relatively weak gas-solid interaction, as exemplified by the adsorption of water and alkanes on nonporous low-polarity solids such as polytetrafluoroethylene (Teflon). In this case, the adsorbate does not effectively spread on the solid surface. Type IV and V isotherms are the characteristic of vapor adsorption reaching an asymptotic value as the saturation pressure is approached. Adsorption of organic vapors on activated carbon is typically type IV, whereas adsorption of water vapor on activated carbon is type V. The shape of the adsorption isotherm of a solute from the solution is sensitive to the competitive adsorption of the solvent and other components and may deviate greatly from that of its vapor on the solid.

As a result, the pore sizes of porous materials can be decided from isotherm types as follows:

- A **type I isotherm** is obtained when $P/P_0 < 1$ and $c > 1$ in the BET equation, where P/P_0 is the partial pressure value and c is the BET constant, which is related to the adsorption energy of the first monolayer and varies from solid to solid. The characterization of microporous materials, those with pore diameters less than 2 nm, gives this type of isotherm.
- A **type II isotherm** is very difficult compared to the Langmuir model. The flatter region in the middle represents the formation of a monolayer. A type II isotherm is obtained when $c > 1$ in the BET equation. This is the most common isotherm obtained when using the BET technique. At very low pressures, the micropores fill with nitrogen gas. At the knee, monolayer formation and multilayer formation occur at a medium pressure. At a higher pressure, capillary condensation occurs.
- A **type III isotherm** is obtained when $c < 1$ and shows the formation of a multilayer. Because there is no asymptote in the curve, no monolayer is formed and BET is not applicable.
- A **type IV isotherm** occurs when capillary condensation occurs. Gases condense in the tiny capillary pores of the solid at pressures below the saturation pressure of the gas. In the lower-pressure regions, it shows the formation of a monolayer followed by the formation of multilayers. The BET surface area characterization of mesoporous materials, which are materials with pore diameters between 2 and 50 nm, gives this type of isotherm.
- **Type V isotherms** are very similar to type IV isotherms and are not applicable to BET.

3.4 Hydrogen Gas Storage

Hydrogen gas storage undoubtedly attracts public interest because it represents a future energy resource based on its high chemical abundance, high energy density, and environmentally friendly characteristics.

The demand of the US Department of Energy (DOE) for hydrogen storage is 5.5 wt% in gravimetric capacity and 40 kg/m^3 in volumetric capacity at an operating temperature of 233–333 K with a pressure of 100 atoms by the year 2017. Recently, a large number of porous materials have been explored for the application of hydrogen storage. However, none of candidates developed so far has satisfied the DOE target. COF materials, as new crystalline porous materials, have also attracted increasing interest in hydrogen storage applications.

The capabilities of hydrogen storage are summarized by using representative COFs in Table 3.2. COF-18 (S_{BET} = 1263 m^2/g, V_P = 0.65 cm^3/g, pore size = 1.8 nm) presents the highest hydrogen uptake at a low pressure (1 bar and 77 K) among similar 2D COFs with different alkyl chain lengths (COF-11Å, COF-14Å, and COF-16Å). For context, 3D COFs (COF-102 and COF-103) showed much higher hydrogen storage capacities because of the higher surface areas and lower densities, relative to 2D COFs. Among the different COFs for H_2 uptake, the one with the largest storage capacity is 3D COF-102 (Table 3.2, S_{BET} = 3620 m^2/g, V_P = 1.55 cm^3/g, pore size = 12 nm), which takes up 72 mg/g at 1 bar and 77 K. This capacity is comparable to those of metal-organic framework 177 (MOF-177, 75 mg/g, S_{BET} = 4500 m^2/g), MOF-5 (76 mg/g, S_{BET} = 3800 m^2/g), and the porous aromatic framework PAF-1 (75 mg/g, S_{BET} = 5600 m^2/g). These capacities significantly demonstrate the potential of COFs for use as hydrogen storage materials.

However, hydrogen storage of COFs at ambient pressure and temperature is still low and far from meeting the DOE requirements. Adsorption simulations of COFs show that doping with charged species can affect their capacity. Theoretical studies show that metal-doped COFs can enhance hydrogen storage at a temperature range of 273 to 298 K [5, 6]. Recently, an improvement in the hydrogen storage capacity has been theoretically predicted by lithium-doped COFs due to the increased binding energy between the H_2 and Li atoms. These simulation studies indicate that hydrogen uptake by Li-doped COF-105 and COF-108 (6.84 and 6.73 wt%, respectively, at 298 K and 100 bar) was more than that by nondoped COFs and previously reported MOF materials. In spite of the fact that experimental and theoretical results indicate that COFs are promising candidates for hydrogen storage, practical and industrial use of COFs toward the DOE target for hydrogen uptake is still far away.

Table 3.2 Selection of COFs with BET, porosity parameters, and H_2 capture

COFs	BET surface area (m^2/g)	Pore size (nm)	Pore volume (V_p, cm^3/g)	H_2 uptake (wt%)
COF-1	750	0.9	0.3	1.48
COF-5	1670	2.7	1.07	3.58
COF-6	750	0.64	0.32	2.26
COF-8	1350	1.87	0.69	3.5
COF-10	1760	3.41	1.44	3.92
COF-102	3620	1.15	1.55	7.24
COF-103	3530	1.25	1.54	7.05
COF-1	628	0.9	0.36	1.28 (1 bar)
COF-11Å	105	1.1	0.05	1.22 (1 bar)
COF-14Å	805	1.4	0.41	1.23 (1 bar)
COF-16Å	753	1.6	0.39	1.4 (1 bar)
COF-18Å	1263	1.8	0.69	1.55 (1 bar)
CTC COF	1710	2.26	1.03	1.12 (0.5 bar)
Dual-pore COF	1771	0.73, 2.52	3.189 (P/P^0)	1.37 (1 bar)

Source: Reproduced from Ref. [4] under the Creative Commons Attribution License (https://creativecommons.org/licenses/by/4.0/).

3.5 Methane Gas Storage

Methane gas stands out as a potential vehicular fuel; however, the lack of an effective, economic, and safe on-board storage system is a major technical barrier that prevents methane-driven automobiles from competing with the traditional ones. According to the US DOE requirements, the current target for methane storage is 350 cm^3_{STP}/cm^3 absorbent (v/v) and 0.5 $g_{CH4}/g_{adsorbent}$ (699 $cm^3/_{STP}g$)

at 35 bar and 298 K (STP: standard temperature and pressure). In Table 3.3, COF-1 as a 2D COF shows CH_4 uptake of 40 mg/g at 298 K and 35 bar. However, upon increasing the pressure, no increase in methane adsorption is observed because of the small pores in COF-1. The pores in COF-1 are quickly occupied by CH_4 molecules at low pressures and higher pressures may not compact the gas any more. COF-102 has the largest pores as a 3D COF and shows a higher CH_4 storage capacity (187 mg/g). COF-103 possesses a high methane uptake capacity (up to 175 mg/g). These values are comparable to the highest methane uptake by an MOF (MOF-210; 210 mg/g). Theoretical simulation studies indicate that lithium-doped COFs could significantly strengthen the building between methane and Li^+ cations, which will further improve the methane capacity of COFs. In these results, methane storage by lithium-doped COFs is double at 298 K and low pressures (>50 bar), as compared to that by nondoped COFs.

The above-mentioned studies indicate the potential application of COFs for methane storage at ambient temperatures and 35 bar. However, the practical use of COFs to meet the requirements of the DOE is a big challenge.

Table 3.3 BET surface area, porosity, parameters, and CH_4 capture of selected COFs

COFs	BET Surface area (m^2/g)	Pore size (nm)	Pore volume (V_P, cm^3/g)	CH_4 uptake (mg/g)
COF-1	750	0.9	0.3	40
COF-5	1670	2.7	1.07	89
COF-6	750	0.64	0.32	65
COF-8	1350	1.87	0.69	87
COF-10	1760	3.41	1.44	80
COF-102	3620	1.15	1.55	187
COF-103	3530	1.25	1.54	175
ILCOF-1	2723	0.23	1.21	129 (L/L)

Source: Reproduced from Ref. [4] under the Creative Commons Attribution License (https://creativecommons.org/licenses/by/4.0/).

3.6 Carbon Dioxide Gas Storage

Carbon dioxide is the primary greenhouse gas that is responsible for global warming, rising sea levels, and the increasing acidity of the oceans. The possibility of capturing CO_2 from industrial emission sources has attracted broad interest. Among the current strategies, adsorption by porous materials is energetically efficient and technically feasible. In 2009, Yaghi and coworkers studied a family of 2D COFs with 1D micropores (COF-1 and COF-6), 2D COFs with mesopores (COF-5, COF-8, and COF-10), and 3D COFs with 3D medium-sized pores (COF-102 and COF-103) on the application of CO_2 capture. Table 3.4 shows the storage of CO_2 at 298 K and 55 bar, in the order COF-102 > COF-103 > COF-10 > COF-5 > COF-8 > COF-6 > COF-1, which indicates that high-pressure CO_2 capture is highly related to the pore volumes and SA_{BET} of COFs.

Functional groups inside the channel walls of COFs can enhance CO_2 adsorption and separation. Fluorinated alkanes exhibit an extraordinary affinity to CO_2 molecules. Han and coworkers reported fluorinated triazine-based COFs as FCFT-1 and FCTF-1-600 for CO_2 capture by using tetrafluoroterephthalonitrile as a monomer [8]. Triazine-based COFs possess N-rich frameworks, which are favored for CO_2 adsorption. Meanwhile, the high electron negativity of F enhances CO_2 electrostatic interaction. The CO_2 adsorption values of these triazine-based COFs at 273 K and 1 bar were as follows: FCTF-1-600 (124 cm^3/g) > FCTF-1 (105 cm^3/g) > CTF-1-600 (86 cm^3/g) > CTF-1 (55 cm^3/g). These results demonstrate that the fluorinated COFs show much better CO_2 uptake under low pressure. Jiang et al. converted a conventional hydroxy group-modified 2D COF ($[HO]_{x\%}$-H_2P COFs) into an outstanding CO_2-capturing material, carboxyl group-modified 2D COF ($[HO_2C]_{x\%}$-H_2P COFs), via the quantitative ring opening reaction of succinic anhydride, as shown in Fig. 3.3 [9]. $[HO_2C]_{x\%}$-H_2P COFs show micropores ranging from 1.4 to 2.2 nm instead of mesopores of $[OH]_{x\%}$-H_2P COFs (2.5 nm). From the results of Table 3.4, the $[OH]_{x\%}$-H_2P COFs have a low CO_2 capacity (between 23 and 32 cm^3/g) at 273 K and 1 bar. However, $[HO_2C]_{x\%}$-H_2P COFs exhibit enhanced CO_2 adsorption between 49 and 89 cm^3/g under the same conditions.

Table 3.4 BET surface area, porosity parameters, and CO_2 capture of selected COFs

COFs	BET surface area (m^2/g)	Pore size (nm)	Pore volume (V_P, cm^3/g)	CO_2 uptake (cm^3/g)		
				Low[a]	High[b]	Ref.
COF-1	750	0.9	0.3	51	117	[7]
COF-5	1670	2.7	1.07	31	443	
COF-6	750	0.64	0.32	85	158	
COF-8	1350	1.87	0.69	33	321	
COF-10	1760	3.41	1.44	27	514	
COF-102	3620	1.15	1.55	34	611	
COF-103	3530	1.25	1.54	38	606	
FCTF-1	662	0.46, 0.54	-	105		[8]
FCTF-1-600	1535	0.46, 0.59	-	124		
CTF-1	746	0.54		55		
CTF-1-600	1553	-		80		
$[HO]_{25\%}$-H_2P COF	1054	2.5		27		
$[HO]_{50\%}$-H_2P COF	1089	2.5		23		
$[HO]_{75\%}$-H_2P COF	1153	2.5		26		
$[HO]_{100\%}$-H_2P COF	1284	2.5		32		

(*Continued*)

Table 3.4 (*Continued*)

COFs	BET surface area (m^2/g)	Pore size (nm)	Pore volume (V_P, cm^3/g)	CO_2 uptake (cm^3/g)		
				Low[a]	High[b]	Ref.
$[HCO_2]_{25\%}$-H_2P COF	786	2.2		49		
$[HCO_2]_{50\%}$-H_2P COF	673	1.9		68		
$[HCO_2]_{75\%}$-H_2P COF	482	1.7		80		
$[HCO_2]_{100\%}$-H_2P COF	364	1.4		89		
$[HC{\equiv}C]_0$-H_2P COF	1474	2.5	0.75	37		[9]
$[HC{\equiv}C]_{25}$-H_2P COF	1413	2.3	0.71	27		
$[HC{\equiv}C]_{50}$-H_2P COF	962	2.1	0.57	24		
$[HC{\equiv}C]_{75}$-H_2P COF	683	1.9	0.42	22		
$[HC{\equiv}C]_{100}$-H_2P COF	462	1.6	0.28	20		
$[Et]_{25}$-H_2P COF	1326	2.2	0.55	28		
$[Et]_{50}$-H_2P COF	821	1.9	0.48	23		
$[Et]_{75}$-H_2P COF	485	1.6	0.34	21		
$[Et]_{100}$-H_2P COF	187	1.5	0.18	19		
$[MeOAc]_{25}$-H_2P COF	1238	2.1	0.51	43		
$[MeOAc]_{50}$-H_2P COF	754	1.8	0.42	45		
$[MeOAc]_{75}$-H_2P COF	472	1.5	0.31	42		

COFs	BET surface area (m^2/g)	Pore size (nm)	Pore volume (V_P, cm^3/g)	CO_2 uptake (cm^3/g)		
				Low[a]	High[b]	Ref.
$[MeOAc]_{100}$-H_2P COF	156	1.1	0.14	33		
$[AcOH]_{25}$-H_2P COF	1252	2.2	0.52	48		
$[AcOH]_{50}$-H_2P COF	866	1.8	0.45	60		
$[AcOH]_{75}$-H_2P COF	402	1.5	0.32	55		
$[AcOH]_{100}$-H_2P COF	186	1.3	0.18	49		
$[EtOH]_{25}$-H_2P COF	1248	2.2	0.56	47		
$[EtOH]_{50}$-H_2P COF	784	1.9	0.43	63		
$[EtOH]_{75}$-H_2P COF	486	1.6	0.36	60		
$[EtOH]_{100}$-H_2P COF	214	1.4	0.19	43		
$[EtNH_2]_{25}$-H_2P COF	1402	2.2	0.58	59		
$[EtNH_2]_{50}$-H_2P COF	1044	1.9	0.5	80		
$[EtNH_2]_{75}$-H_2P COF	568	1.6	0.36	68		
$[EtNH_2]_{100}$-H_2P COF	382	1.3	0.21	49		

[a]CO_2 uptake was measured at 1 bar and 273 K.

[b]CO_2 uptake was measured at 55 bar and 298 K.

Source: Reproduced from Ref. [4] under the Creative Commons Attribution License (https://creativecommons.org/licenses/by/4.0/).

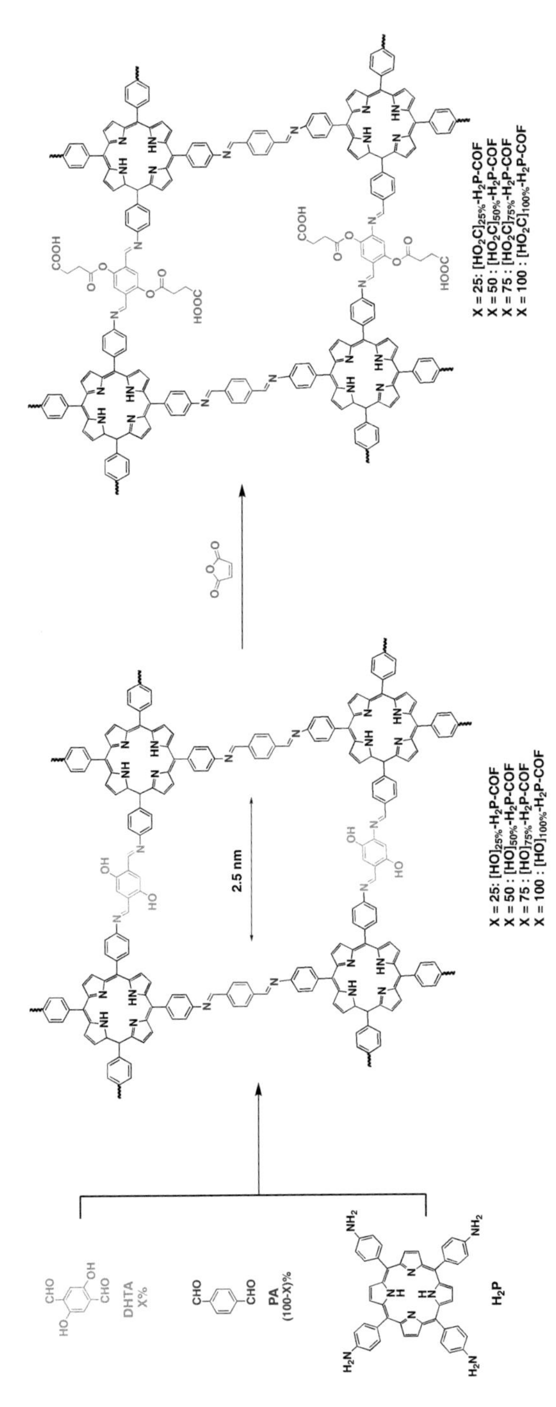

Figure 3.3 Synthesis of $[OH]_{X\%}$-H_2P COFs with channel walls functionalized with carboxylic acid groups through the ring-opening reaction of $[OH]_{X\%}$-H_2P COFs with succinic anhydride.

Jiang et al. also reported a similar strategy to modify the pore surface with different functional groups via a click reaction [10]. In Table 3.4, it's clear that different functions of pore walls of COFs can induce a decrease in surface areas, pore sizes, and pore volumes. However, the capacity of CO_2 adsorption is highly dependent on the structures of the functional groups. Ethyl-functionalized COFs, $[Et]_x$-H_2P COFs, show a CO_2 adsorption capacity similar to those of $[CH{\equiv}C]_x$-H_2P COFs. The CO_2 adsorption capacities of $[CH{\equiv}C]_x$-H_2P COFs and $[EtNH_2]_x$-H_2P COFs is enhanced with an increase in the BET surface area and pore size. However, $[MeOAc]_x$-H_2P COFs, $[AcOH]_x$-H_2P COFs, $[EtOH]_x$-H_2P COFs, and $[EtNH_2]_x$-H_2P COFs exhibit much higher CO_2 adsorption capacities than those of $[CH{\equiv}C]_x$-H_2P COFs. The adsorption capacities are in the order of $[CH{\equiv}C]_x$-H_2P COFs $\approx$ $[Et]_x$-H_2P COFs $\ll$ $[MeOAc]_x$-H_2P COFs $<$ $[AcOH]_x$-H_2P COFs $\approx$ $[EtNH_2]_x$-H_2P COFs, which is because of the interaction between the functional group and the CO_2 molecule. As these results show, the CO_2 adsorption performance is highly related to the pore surface of the COFs.

3.7 Membrane Separation of COFs

The implementation of the membrane technology in the context of resource recovery and sustainable development has demonstrated an ecofriendly potential for overcoming energy and environmental challenges. Due to its low energy consumption, small carbon footprint, and easy operation and scalability, membrane separation has experienced rapid growth in the past few decades. Until now, a variety of membranes with pore sizes ranging from several micrometers to subnanometers have been fabricated from conventional amorphous polymers, such as polyimide, polyamide (PA), polyacrylonitrile, polyether sulfone (PES), polysulfone, polyvinylidene fluoride (PVDF), and poly(amide-imide)-based cross-linked polymers [11]. However, these polymers are not optimal for transport because the resulting membranes lack ordered and tunable pore channels, leading to a typical nonuniform pore size, limited porosity, and poor interconnectivity. Novel membranes fabricated from porous crystalline materials with uniform pore channels are significantly advanced in selectivity and porosity.

Crystalline materials such as zeolites and MOFs have demonstrated a superior performance over conventional polymer membranes. However, due to their relatively low adhesion to a polymeric support, partly organic nature, and possible defects between crystals, the full capability of MOF-based membranes is relatively difficult to realize. Therefore, the development and design of new porous crystalline polymers with ordered pore structures and tunable pore sizes are important for the synthesis of advanced separation membranes.

Up to now, studies on COF-based membranes have cover a broad range of topics, including gas separation membranes, separation membranes for liquid phase, and proton exchange membranes in fuel cells. This section is an introduction and discussion of membrane design strategies based on COFs. The optimal method to synthesize a high-performance COF membrane in terms of high perm-selectivity will be highlighted. Especially, this section summarizes the applications of COF membranes in gas separation.

3.7.1 Key Properties of COFs for Membrane Separation

Due to their well-defined and precise pore structure with relatively large surface areas, COFs were first used for gas storage. Recently, along with the emergence of well-designed and multifunctional COFs, their application has gradually extended to membrane technology for efficient and precise separation. Different from gas storage, membrane separation covers a wide variety of areas, including gas separation, water purification/treatment, organic solvent nanofiltration, pervaporation, and fuel cells, in which the membranes are exposed to various mild/harsh environments. This makes the synthesis of suitable COFs for membrane separation more complicated compared to other applications, requiring rational selection criteria for fabricating COF-based membranes. Specifically, the physical pore size defines the application of membranes, for example, in reverse osmosis, nanofiltration, ultrafiltration, and microfiltration. So the selection of COFs with specific pore sizes is key for membrane design. In addition, membrane fabrication uses mild or harsh solvents and mechanical treatments; thus, the chemical and mechanical stability of COFs is another selection criterion. Other properties, like hydrophilicity/hydrophobicity, surface charge, and proton conductivity, also have a significant influence on the performance of COF-based membranes in various separations. Altogether, the selection of COFs has many variables that determine the separation performance and future applications in membrane technology. The best way to construct desired COF-based membranes with a high perm-selectivity and long-term stability is the rational selection of COFs on the basis of various combinations of their properties.

3.7.2 Fabrication of COF-Based Membranes

3.7.2.1 Design principles

A rational design strategy bridging the gap between COFs and COF-based membranes is the key to optimizing the advantages of COFs for precise and rapid membrane separation. When compared with other porous materials, COFs have orderly and closely arranged pore structures, which is the main advantage of COFs in membrane applications. Therefore, how to achieve the full potential of these orderly arranged pore structures is the core challenge in fabricating a COF-based membrane. Initially, COFs were blended into polymeric matrix membranes as porous fillers to gain additional channels for gas, water, and solvents to pass through. However, the pore structures resulting from the phase inversion of polymer solutions were still the dominating transport pathways. With advancing membrane fabrication techniques and COF synthesis methods, the pore structures in COFs are gaining increasing significance in membrane separation. Continuous COF-based membranes were synthesized from uninterrupted and pure COFs by in situ growth, layer-by-layer stacking, and interfacial polymerization (IP). These methods will be discussed in detail next.

3.7.2.2 Blending

Integrating COFs into common membrane fabrication methods, including nonsolvent-induced phase inversion (NIPS) and IP, is a simple and reproducible approach for fabricating COF-based hybrid membranes. Compared to other classical inorganic particles, the fully organic nature endows COFs with excellent compatibility with polymer matrices. Moreover, the pore structures of COFs provide additional passages for gas, water, and organic solvent molecules to permeate and enhance the perm-membrane fabrication methods. The routes of introducing COFs in these two approaches are different. In a typical NIPS, COFs are directly added in the organic solvents with polymers to achieve a homogeneous dope solution. Then a film of a certain thickness is formed by casing the solution on a glass plate or nonwoven and subsequently experiences an induced phase inversion while being immersed in water, to form a membrane. The final COF-based membrane is obtained after the removal of the leftover solvent by immersion in deionized water. These COF-based membranes are also known as COF-based mixed matrix membranes (MMMs). In the IP process, the incorporation of COFs in a thin PA film is realized by dispersing COFs with the diamine monomers in the water phase. During the reaction between diamine monomers from the aqueous phase and the acid chloride monomers from organic phase, the COFs dispersed in the aqueous phase are trapped in the formed active thin layer. These two approaches of blending COFs in polymer membranes will be highlighted in this section.

In the last three years, various MMMs have been constructed by embedding COFs, such as SNW-1, COF-1, and functional COFs, in polymers. In particular, 2D SNW-1 has been extensively investigated because of its relatively low synthesis cost and high hydrophilicity. Pioneering SNW-1-based membranes were fabricated by incorporating COF SNW-1 particles with a diameter of 50–70 nm into sodium alginate (SA) matrices for ethanol dehydration [12]. The incorporation of SNW-1 particles endowed the membranes with an enhanced thermal/mechanical stability and good antiswelling properties. The SA-SNW-1 (25) membrane with an SNW-1 loading of 25 wt% achieved the optimal separation performance with a permeation flux of 2397 g/m^2 h and a separation factor of 1293 for ethanol dehydration, which was superior to that of other porous material–based membranes due to the porous structure of SNW-1.

In addition to SNW-1, ketoenamine-linked COFs, TpPa-1, TpPa-2, TpHZ, TpBD, NUS-10, and NUS-9, have received tremendous attention in synthesizing COF-based MMMs because of their relatively high stability and hydrophilicity. Ketoenamine-linked COFs shaped like hollow spheres, particles, and nanosheets were explored in membrane fabrication and separation. Spherical flower-like TpBD and TpPa-1 particles were used as the active phase in the polymer (polybenzimidazole–5-*tert*-butylisophthalic acid [PBI-BuI]) for gas separation (Fig. 3.4) [13]. The compatibility between COFs and PBI-BuI was significantly improved by creating intermolecular interactions between H-bonded benzimidazole groups of PBI and COFs; thus a considerably higher loading of COF (50%) in the polymer matrix was achieved. The obtained membrane with a loading of 50% TpBD showed a sevenfold elevation in the permeability of gases (H_2, N_2, CH_4, and CO_2) compared with the pristine membrane, with small deviations in selectivity. The improved permeation was mainly due to the relatively low diffusion resistance achieved by incorporated COFs, while the selectivity was still governed and maintained by the polymer matrix. Assisted by a similar strategy, in 2018, TpBD with hollow nanosphere structures was constructed by a facile template-directed approach followed by mixing with SA matrices to produce membranes for dehydration [14]. The structure of hollow nanospheres was realized by etching the Fe_3O_4 core of TpBD@Fe_3O_4 obtained through the synthesis of TpBD on the template of an Fe_3O_4 nanocluster, in a HCl solution (Fig. 3.5). The incorporation of hollow TpBD nanospheres into the polymer matrix allowed the prepared membrane to have improved hydrophilicity and a large number of microporous channels resulting from the pore structure of hollow nanospheres and the free volume cavities between COFs and the polymer. In addition, due to the highly porous structure of hollow TpBD nanospheres, the porosity of TpBD-based MMMs was notably boosted, resulting in a low flow resistance. COF nanosheets peeled off from the bulk COF particles were also investigated for fabricating MMMs. Cao et al. embedded 2D TpHz nanosheets with a lateral size of 200–400 nm and a thickness of 4.5 nm in a PES matrix to fabricate functional MMMs [15]. An interesting feature in this study was that the content of the COFs on the membrane surface can be up to 50.90 vol% although only 12 wt% of TpHz was added in the dope solution; this was primarily ascribed to the migration of COF nanosheets from the matrix to the surface during the membrane formation.

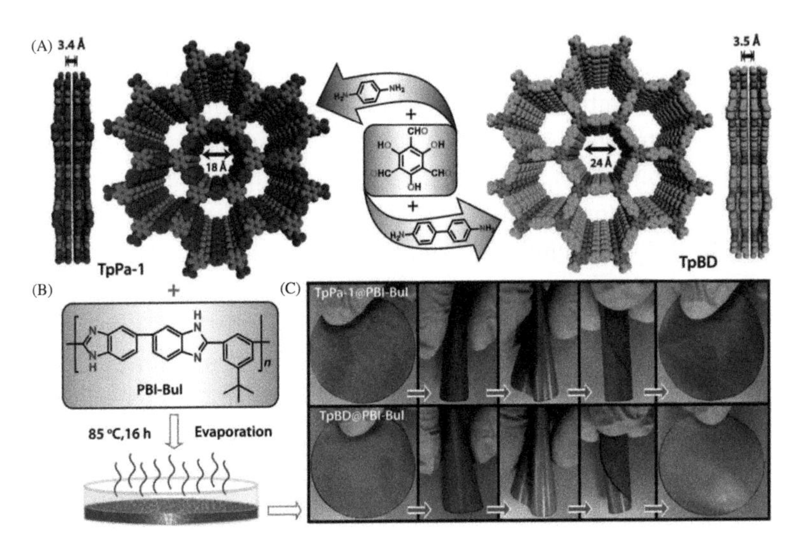

Figure 3.4 COFs/PBI-Bul membrane. (A) Schematic representations of the synthesis of COFs and their packing models, indicating the pore aperture and stacking distances. (B) Overview of the solution-casting method for COF@PBI-Bul hybrid membrane fabrication. (C) Digital photographs showing the flexibility of TpPa-1 and TpBD(50)@PBI-Bul. Reproduced with permission from Ref. [13]. Copyright (2016), John Wiley and Sons.

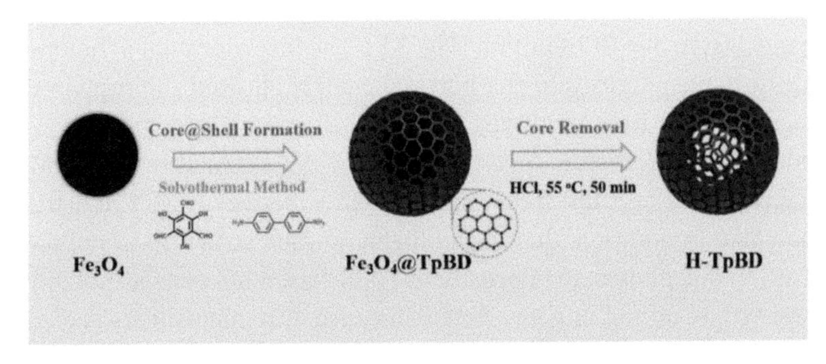

Figure 3.5 The synthesis protocols and chemical structures of the H-TpBD. Reprinted from Ref. [14], Copyright (2018), with permission from Elsevier.

Compared with pristine COFs, functionalized COFs allow for the fabrication of versatile and tailored membranes with diverse properties. Dan et al. reported the synthesis of COFs (NUS-9 and NUS-10) with an intrinsic proton-conductive performance by introducing sulfonic acid groups in organic linkers [16]. These two functional COFs were then incorporated in PVDF to prepare proton exchange membranes.

Unlike MMMs, TFN membranes are synthesized by a rapid IP of two monomers from the organic and aqueous phase. The viscosity of the aqueous phase is much lower than that of the dope solution for the synthesis of MMMs. Thus achieving a good dispersion of COFs in solution is much easier and not time consuming. Besides, COFs are only incorporated in the thin layer of TFN membranes, which would result in a relatively low consumption of COFs during membrane fabrication. The first COF-incorporated TFN membrane was reported by Wu et al. in 2016 [17]. In their study, SNW-1 particles bearing secondary amine groups were introduced in the PA thin film via IP of trimesoylchloride (TMC) and piperazine monomers on top of a PES substrate. The –NH– groups of SNW-1 are involved in IP with –COCl groups of TMC, so the interface compatibility between SNW-1 and PA is thought to be high, which results in the high stability of SNW-1 in the thin active film. The obtained COF-based TFN membranes exhibited an increased pure water flux from 10 L m^{-2} h^{-1} bar^{-1} to 19.25 m^{-2} h^{-1} bar^{-1} while maintaining a Na_2SO_4 rejection of over 80%. The enhanced performance was mainly attributed to the improved hydrophilicity and the increased number of water channels derived from the pore structure of SNW-1 and the voids at the interface between SNW-1 and the polymer. Similarly, the same COF, SNW-1, was embedded in a TFN membrane synthesized via IP of *m*-phenylenediamine and TMC (Fig. 3.6) for organic solvent nanofiltration [18]. The resultant membranes demonstrated an improved surface hydrophilicity and a decreased skin layer thickness, and thus a 46.7% increment of ethanol permeation and an increased rhodamine B (476 Da) rejection of up to 99.4% were achieved compared with COF-free membranes. Currently, although only SNW-1 has been applied in fabricating COF-based TFN membranes, many more different COFs can be explored in fabricating COF-based TFN membranes due to their specific characteristics, such as porous structures arranged in an orderly manner and relatively low density. For instance, chemically stable COFs with pore sizes of around 1 nm—TpHz, ACOF-1, COF-300, TR-CIF-1, CTF-1, and COF-LIZ1—also have great potential to be applied in synthesizing TFN membranes with relatively high separation performance, although they have not been explored yet.

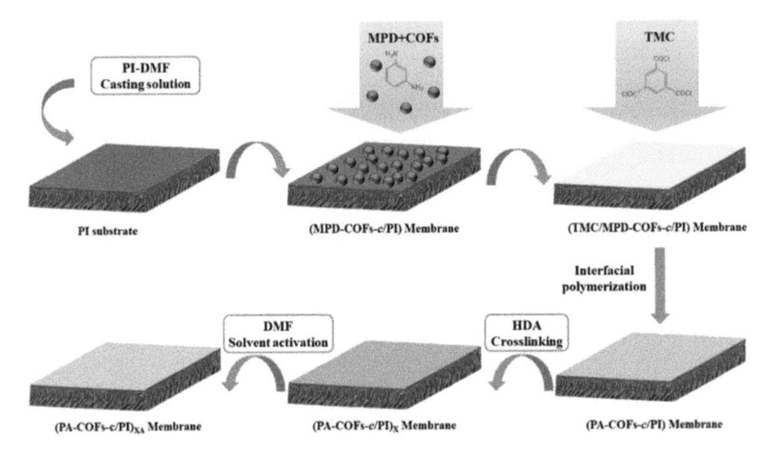

Figure 3.6 Schematic illustration of the fabrication of the TFN-OSN membrane with COFs incorporated inside the Paskin layer. Reprinted from Ref. [18], Copyright (2019), with permission from Elsevier.

3.7.2.3 In situ growth

Generally, discontinuous COF-based membranes cannot fulfill the potential of the pure structures of COFs for membrane separation. Thus, advanced membranes with high perm-selectivity are difficult to synthesize by blending. Therefore, establishing methods of preparing continuous COF-based membranes is highly necessary for obtaining a higher separation performance. With advances in COF synthesis, continuous COF membranes were grown in situ by refining the current COF synthesis methods. Both freestanding COF films and COF-based composite membranes can be designed by in situ growth.

Kharul et al. reported that in situ growth of freestanding COF-based membranes was pioneered in 2017 (Fig. 3.7) [19]. Briefly, in the first step, the powered aromatic diamine the coreagent, *p*-toluene sulfonic acid (PTSA), and water (PTSA·H_2O) were mixed in water to form an organic salt. The resultant salts and Tp were shaken thoroughly using a vortex shaker to make the dough. Then, it was knife-cast on a clean glass plate to generate a film. Finally, COF-based membranes were produced via the in situ growth of COFs by baking the film at 60°C–120°C in a programmable oven for 12–72 h. The obtained continuous COF-based membrane (M-TpTD) showed an acetonitrile flux of 260 L m^{-2} h^{-1} bar^{-1}, which was 2.5 times higher in magnitude than the PA-based literature-reported NF membranes with similar solute rejection (~99%). This relatively high perm-selectivity was mainly because the pores of COFs are fully employed as the channels for solvent molecules to be transported. This strategy was further extended to synthesize TpBD$(Me)_2$-, TpAzo-, and TpBpy-based proton exchange membranes by slightly changing the casting of the reactant mixture on a mold instead of a glass plate [20]. More recently, TpOMe-Pa1-, Tp-OMe-BD$(NO_2)_2$-, TpOMe-Azo-, and TpOMe-Bpy-based COF membranes were synthesized [21].

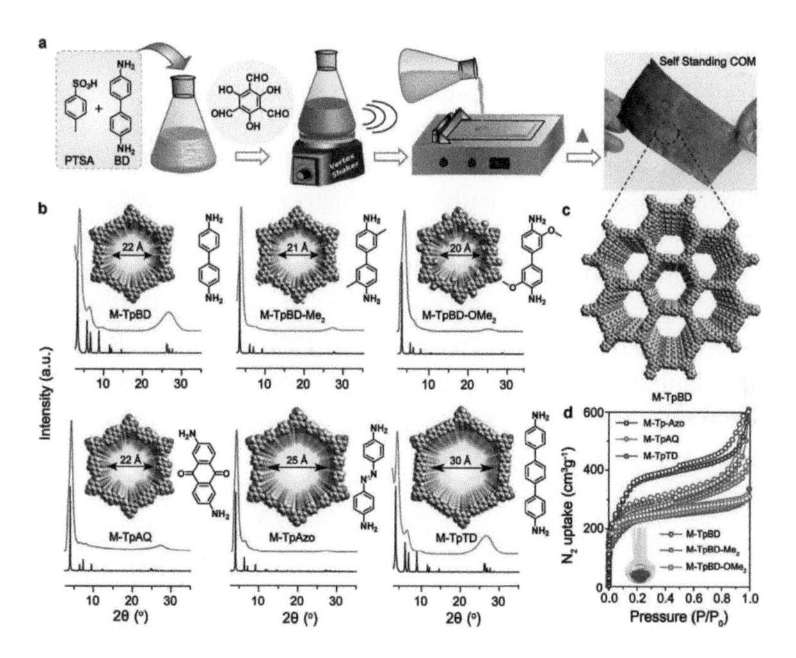

Figure 3.7 In situ growth of COF membranes (COMs). (a) Schematic representation of COM (M-TpBD) fabrication. (b) The pore apertures shown inside the space-filling COM model in the panel have been calculated by measuring the distance between the two opposite arms of the hexagon after subtraction of van der Waals radius of the hydrogen atom. (c) Space-filling packing model of M-TpBD hexagonal frameworks. (d) Comparison of N_2 adsorption isotherms of all six COMs prepared in this work. Reproduced with permission from Ref. [19]. Copyright (2017), John Wiley and Sons.

To address the issue of the low mechanical strength of the freestanding COF-based membranes, COFs were grown in situ on modified or unmodified porous supporting membranes. Taking advantage of this strategy, Caro et al. fabricated continuous and high-quality COF-LUZ1 membranes supported on commercial ceramic tube membranes for dye removal from water [22]. In their research, the surface of the alumina tube was first modified with 3-amino-propyltriethoxysilane and was then further functionalized with aldehyde groups by reaction with 1,3,5-triformylbenzene (TFB) at 150°C for 1 h (Fig. 3.8). The final well-intergrown COF-LZU1 layer on the alumina tube was realized by vertically placing the modified alumina tube in the Teflon-lined stainless-steel autoclave with reactant mixtures and the following reaction at 120°C for 72 h. This strategy was further extended to prepare continuous COFs bilayers on a flat alumina porous substrate [23]. COF-LUZ1-ACOF-1 membranes were fabricated by first synthesizing COF-LUZ1 by condensation of TFB with *p*-phenylenediamine (PDA) at room temperature and then synthesizing high-crystallinity ACOF-1 by condensation of TFB with hydrazine hydrate at a higher temperature. The resultant COF-LUZ1-ACOF-1 bilayer membrane showed a much higher separation selectivity for H_2/CO_2, H_2/N_2, and H_2/CH_4 gas mixture than the individual pore networks, surpassing the Robeson upper bounds. By utilizing a similar strategy, COF-MOF composite membranes were also fabricated [24].

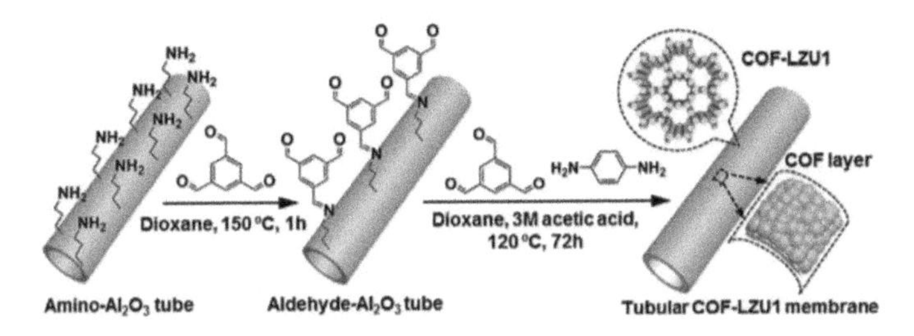

Figure 3.8 Synthesis of tubular COF-LZU1 membranes. Reproduced with permission from Ref. [22]. Copyright (2018), John Wiley and Sons.

3.7.2.4 Layer-by-layer stacking

The route of membrane fabrication via layer-by-layer stacking of nanosheets or monolayers originally came from the synthesis of membranes from graphene and graphene oxide monolayers [25].

This involved exfoliation of bulk materials comprising monolayers in water or solvents. Then, the nanosheets were stacked on a porous substrate by pressure- or vacuum-assisted filtration, or dip coating, to form a continuous thin membrane. This approach is applicable to prepare continuous COF-based membranes from COF nanosheets. Most of the prepared COFs have a 2D lamellar structure, resulting from the planar organic linkers, and 2D COFs nanosheets were easily obtained via the splitting of bulk COFs by breaking the van der Waals force between the neighboring layers. Many different methods of COF exfoliation have been developed, which have been summarized by Tang et al. [26]. Different from the in situ growth approach, the selective COF layer of the synthesized membranes is thinner. This thin layer helps to achieve a low flow resistance when gases, water, and solvents pass through the membrane. The thickness of this layer is controllable by simply varying the amount of COF nanosheets. Tsuru et al. fabricated an ultrathin COF membrane (~100 nm) for gas separation by repeatedly dip-coating COF nanosheets on the support membrane and then drying them at room temperature (Fig. 3.9) [27]. Due to relatively low flow resistance and high porosity, the synthesized membrane showed a high H_2 permeance (1×10^{-6} mol m^{-2} Pa^{-1} s^{-1}), which was approximately 1 magnitude higher than that of most reported MOF membranes. By using this approach, COF/graphene oxide composite membranes for H_2/CO_2 separation and functionalized COF-based membranes for ion sieving were developed [28, 29].

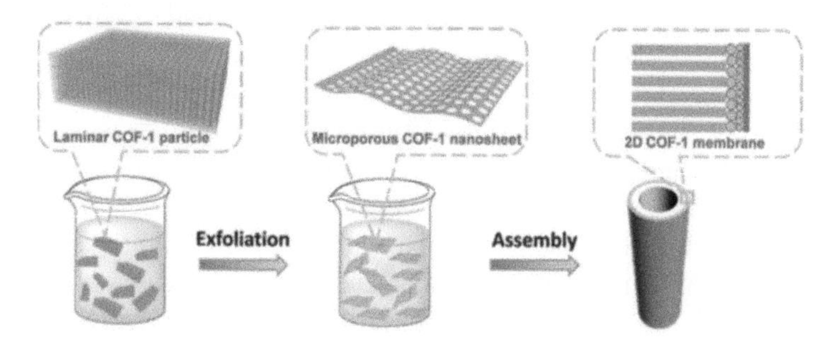

Figure 3.9 Schematic illustration of the preparation of a COF-1 membrane via the assembly of exfoliated COF-1 nanosheets. Reprinted with permission from Ref. [27]. Copyright (2017) American Chemical Society.

3.7.2.5 Interfacial polymerization

IP dominates the synthesis of thin film composite membranes with PA active layers because of its scalability for commercial production and capacity to produce thin separating layers with relatively high water permeability. To achieve the full advantage of the ordered and nanosized pore channels (1–2 nm) of COFs, this method was well adapted to fabricate thin continuous COF-based membranes. In 2017, a pioneering fabrication of continuous COF-based membranes using IP was reported by Banerjee et al. [30]. In this work, the aldehyde organic linker Tp was dissolved in dichloromethane while amine monomers were dissolved in water. The thin active layer was formed by the polycondensation of Tp with diamine and triamine at the interface of dichloromethane and water (Fig. 3.10). To avoid the formation of amorphous polymers during the organic Schiff base reaction, amines were first salt-mediated by PTSA. The formed thin active layer was then transferred on porous grids for filtration measurements. The H bonding in the PTSA-amine decreased the diffusion rate of amine organic linkers, and the reaction rate was slowed down with thermodynamically controlled crystallization. Due to the highly porous structure, the TpBpy thin film membrane displayed unprecedented acetonitrile performance of porous $339\ L\ m^{-2}\ h^{-1}\ bar^{-1}$.

More recently, different from the traditional IP, a new approach of confirming the polymerization of monomers at the oil/water interface by using $Sc(OTf)_3$ as the Lewis acid catalyst was developed [31]. Specifically, the catalyst was preferentially dissolved in water while the monomers terephthalaldehyde (PDA) and 1,3,5-tris(4-aminophenyl)-benzene were dissolved in an organic phase. The reaction was initialized by placing the oily organic solution on the surface of the water phase containing catalysts and freestanding COF-based membranes were formed at the interface.

Apart from the IP that occurred at the interface of liquid/liquid, IP at the liquid/air interface was also established for the fabrication of a continuous COF-based membrane. To achieve the polymerization of monomers at the water/air interface, Lai et al. first synthesized an amphiphilic diamine monomer, 9,9-dihexylfluorene-2,7-diamine (DHF), by attaching two hexyl groups to 2,7-diaminofluorene [32]. Then, a layer of DHF and Tp was formed on the surface of water, followed by the complete evaporation of toluene. The polymerization of monomers at the interface was initiated by adding trifluoroacetic acid to the water. The final yellow TPF-DHF thin membranes with a thickness of 3 nm were obtained after a reaction of 48 h at room temperature.

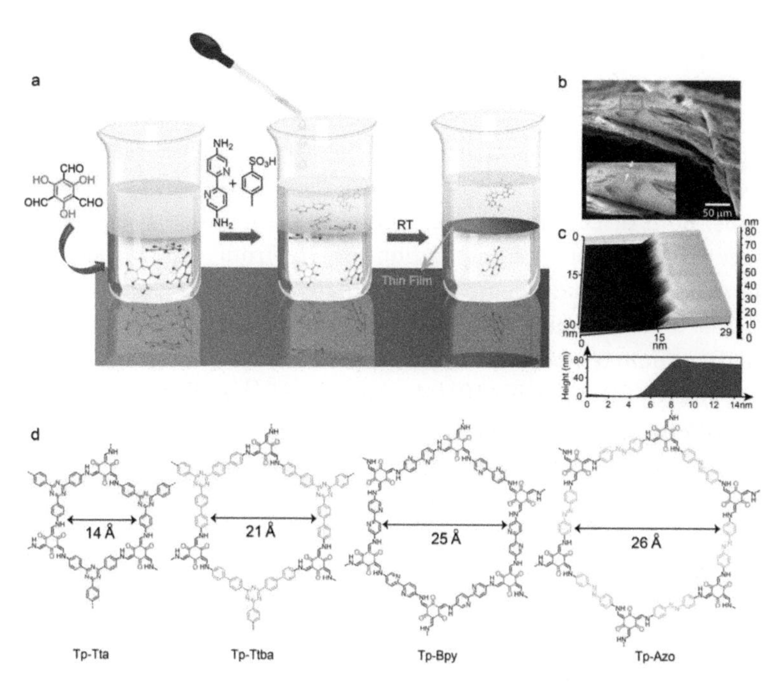

Figure 3.10 Synthesis scheme of COF thin films. (a) Schematic representation of the interfacial crystallization process used to synthesize a TpBpy thin film. The colorless bottom layer corresponds to aldehyde in a dichloromethane solution, the blue layer contains only water as the spacer solution, and the yellow top layer is the Bpy amine-PTSA aqueous solution. (b, c) SEM and AFM images (with corresponding height profiles) of the TpBpy thin film synthesized as illustrated in (a). (d) Chemdraw structures of all the COFs used for synthesizing the thin films via an interfacial crystallization process. Reprinted with permission from Ref. [30]. Copyright (2017) American Chemical Society.

3.7.3 Gas Separation of COF-Based Membranes

Conventional membranes fabricated from amorphous polymers suffer from a disordered and inconsistent pore size and have difficulties in achieving a perm-selectivity surpassing the current

Robeson upper bounds. COFs with abundant and well-ordered in-plane pores are particularly promising in realizing ultrafast and highly selective molecular sieving. Therefore, COF-based membranes for gas separation have been fabricated. The applications of COF-based membranes in gas separation are summarized with particular attention to the separation of H_2 from other gases and CO_2/CH_4 separation due to the rise in energy consumption and because environmental concerns result in demands for clean energy.

Hydrogen is an alternative clean energy source for the replacement of conventional fuels in cars, but the current H_2 production methods inevitably require the separation of unwanted gases. Both discontinuous and continuous COF-based membranes were applied for the separation of H_2 from other gases. A computational study demonstrated that the monolayer COF(CTF-0)-based membrane can achieve an exceptionally high separation factor at room temperature for the separation of H_2/CO_2, H_2/N_2, H_2/CO_2, and H_2/CH_4 up to 9×10^{13}, 4×10^{24}, 1×10^{22}, and 2×10^{36}, which theoretically proved the advantages of COF-based membranes in gas separation. However, the highest separation factor for separation of H_2/CO_2, H_2/N_2, H_2/CH_4 obtained from experimental studies were 31.4 [33], 140 [13], and 84 [23], respectively, which is much lower than the theoretical value. This might be because the experimental COF-based membranes are much thicker than the ideal continuous COF membrane of one monolayer. It is worth pointing out that continuous COF-based membranes normally have a much higher H_2, a relatively high porosity, and low resistance. For example, the highest H_2 permeability (1.7×10^{-6} mol m^{-2} s^{-1} Pa^{-1}) was realized by an ultrathin continuous 2D-CTF-1 membrane (100 nm) [28]. The selectivity of H_2/CO_2 improved with the increase of membrane thickness while the gas permeance reduced.

CH_4 is another attractive substitute for petroleum due to its abundant natural reserves and economic advantages. However, the presence of CO_2 in natural gas leads to a reduced heat value and pipeline corrosion, and efficient removal of CO_2 from CH_2/CH_4 is highly desirable. To select COFs for CO_2/CH_4 separation, Zhong et al. conducted high-throughput computational screening and design of COFs (Fig. 3.11) [34] and 298 COFs with pore-limiting diameters (PLDs) larger than 3.3 Å were evaluated. The membrane selectivity generally increased with the decline of the PLD because of the size exclusion effect. COFs with extremely small PLDs (3.3 Å < PLD < 3.8 Å; the kinetic diameter of CH_4 is 3.8 Å), such as 2D COFs in staggered stacking modes and 3D COFs with interpenetrating configurations, displayed a relatively good performance in CO_2/CH_4 separation. The thermodynamic factor also plays a significant role in the separation performance of COFs. An interaction between the functional sites of COFs and CO_2 would enhance the selectivity between CO_2 and CH_4. For example, COFs with CO_2 favorable interaction sites, like TpMA with a hydroxy group (–OH), TpPa-NO_2 with a nitro group (–NO_2), and ATFGCOF and NUS-2 with a carbonyl group (–C=O–), showed a great potential in CO_2/CH_4, and COFs with 10 different functional group were synthesized. It was found that –NH_2 and –CH_3 contribute little to improving the CO_2/CH_4 separation efficiency. In contrast, 13 COFs with decoration of –F and –Cl are dominant in the top 26 of the modified COFs with CO_2/CH_4 separation factors over 20 because of the great adsorption selectivity.

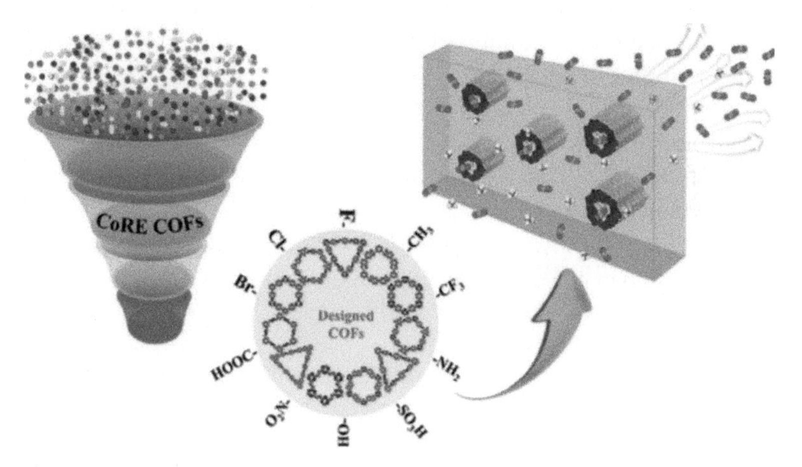

Figure 3.11 Screening and design of COF membranes for CO_2/CH_4 separation. Reprinted with permission from Ref. [34]. Copyright (2019) American Chemical Society.

Experimental studies were also performed on CO_2/CH_4 separation by COF-based membranes [35, 36]. In 2016, COF-based MMMs were prepared by the incorporation of ACOF-1 in the commercial polymer Matrimid 5218 for CO_2/CH_4 separation. Compared with the pure polymer, MMM with 16 wt% of ACOF-1 exhibited an increased CO_2 permeability (5.01×10^{-13} mol m^{-2} s^{-1} Pa^{-1}) with a slight increase of selectivity (~32). The improved permeability was attributed to the additional pathways introduced by the porous ACOF-1. More recently, ACOF-1 was also in situ grown on a porous a-Al_2O_3 substrate for CO_2/CH_4 separation [35]. This continuous COF-based membrane displayed a good thermal stability and superhigh values of both permeability (9.9×10^{-9} mol m^{-2} s^{-1} Pa^{-1}) and selectivity (86.4). The overall performance surpassed the latest Robeson upper bounds for the CO_2/CH_4 gas pair. The high performance is attributed to the excellent CO_2 adsorption capacity of ACOF-1 and the stacked pores with narrower aperture sizes. From these two works, it is obvious that continuous COF-based membranes are superior in CO_2/CH_4 separation compared with discontinuous membranes (COF-Based MMMs) because the advantages of the pore structures of COFs are fulfilled in separation by the continuous membrane.

CO_2/N_2 separation has also attracted tremendous attention to the capture of carbon dioxide to address the issue of global warming. The pioneering separation of CO_2/N_2 by COF-based membranes was conducted by a computational study in 2016 [37]. This study concluded that the interlayer passages formed between the stacked nanosheets have a "gate closing" effect on the selective transport performance of the ultrathin COF-based membranes. Tuning the stacking modes of COF nanosheets resulted in a high permeability of CO_2 and a good CO_2/N_2 selectivity. In 2017, an experimental study of CO_2/N_2 separation by COF-based MMMs showed that the incorporation of SNW-1 in polymers of intrinsic microporosity could enhance the permeability (2.5×10^{-10} mol m^{-2} s^{-1} Pa^{-1}) of CO_2 and selectivity (22.5) of CO_2/N_2 [38].

On the basis of these highlighted experimental studies and the applications of COFs, it can be derived that COFs with pore sizes of less than or around 1 nm, such as CTF-0, CTF-1, ACOF-1, NUS-2, COF-320, and SNW-1, have potential in gas separation, with high selectivities. This indicates that pore size is one of the key factors determining the permeability and selectivity of a COF-based membrane, which was proved by a computational study [34]. Therefore, further development in COF-based membranes for gas separation relies on the synthesis of new COFs with relatively small pore sizes. In addition, although simulation studies reveal that the functional groups on COFs have a significant influence on the performance of gas separation, no experimental study has explored this to date.

3.8 Outlook and Conclusions

COFs, which are constructed from organic linkers, are a new class of crystalline porous materials comprising periodically extended and covalently bound network structures. The intrinsic structures and tailorable organic linkers endow COFs with low densities, large surface areas, tunable pore sizes and structures, and facilely tailored functionality, attracting increasing interest from a lot of different fields. In this chapter, gas adsorption and storage of COFs were described, from the principle of gas adsorption to the separation of gases. Especially, this chapter focused on gas adsorption, storage, and separation for COF-based membranes. To apply COFs in membrane separation, various strategies, including blending, in situ growth, layer-by-layer stacking, and IP, have been established to synthesize COF-based membranes. These membranes show a relatively high performance in gas separation, displaying a huge potential in membrane technology. It is believed that COF-based membranes will be intensively studies in the field of membrane technology for years to come. Specifically, with the advances in separation, other properties of COFs, including polarity, gas favorable properties, and hydrophobicity, are also expected to improve for organic solvent nanofiltration and gas separation, and a combination of these properties on one COF would be a breakthrough in fabricating multifunctional COF-based membranes. In addition, compared to blending, in situ growth, layer-by-layer stacking, and IP result in continuous COF-based membranes, achieving the full capability of COFs in separation. In particular, ultrathin COF-based membranes can be synthesized from layer-by-layer stacking and IP, demonstrating superhigh permeability, making these two methods promising for the future development of COF-based membranes.

Challenges still remain for the future development and application of COF-based membranes in academia and industrial sectors. COFs with pore sizes less than 1 nm are required to construct gas separation membranes with high selectivity by using methods such as blending, in situ growth, layer-by-layer stacking, and IP. However, there is a limitation to these synthetic methods. A further narrowing of the pore sizes of COFs might be achieved by anchoring side groups in the inner walls of COFs via postmodification. Finally, the relatively high cost of COFs and the complicated and time-consuming fabrication methods might hinder large-scale fabrication of membranes. Therefore, low-cost COFs and cost-effective fabrication strategies should be developed to control the cost of COF-based membranes in industrial applications.

References

1. Furukawa, H., Yaghi, O. M. (2009). *J. Am. Chem. Soc.*, **131**, 8875–8883.
2. Taylor, H. S. (1931). *J. Am. Chem. Soc.*, **53**, 578–597.
3. Langmuir, I. (1916). *J. Am. Chem. Soc.*, **38**, 2221–2295.
4. Langmuir, I. (1918). *J. Am. Chem. Soc.*, **40**, 1361–1403.
5. Cao, D., Lan, J., Wang, W., Smit, B. (2009). *Angew. Chem. Int. Ed.*, **48**, 4730–4733.
6. Klontzas, E., Tylianakis, E., Froudakis, G. E. (2009). *J. Phys. Chem. C*, **113**, 21253–21257.
7. Uribe-Romo, F. J., Doonan, C. J., Furukawa, H., Oisaki, K., Yaghi, O. M. (2011). *J. Am. Chem. Soc.*, **1338**, 11478–11481.
8. Zhao, Y., Yao, K. X., Teng, B., Zhang, T., Han, Y. A. (2013). *Energy Environ. Sci.*, **6**, 3684–3692.
9. Huang, N., Chen, X., Krishna, R., Jiang, D. (2015). *Angew. Chem. Int. Ed.*, **54**, 2686–2990.
10. Huang, N., Krishna, R., Jiang, D. (2015). *J. Am. Chem. Soc.*, **137**, 7079–7082.
11. Karan, S., Jiang Z., Livingston, A. G. (2015). *Science*, **348**, 1347–1351.
12. Yang, H., Wu, H., Pan, F., Li, Z., Ding, H., Liu, G., Jiang, Z., Zhang, P., Cao, X., Wang, B. (2016). *J. Membr. Sci.*, **520**, 583–595.
13. Biswal, B. P., Chaudhari, H. D., Banerjee R., Kharul, U. K. (2016). *Chem.-Eur. J.*, **22**, 4695–4699.
14. Yang, H., Cheng, X., Cheng, X., Pan, F., Wu, H., Liu, G., Song, Y., Cao, X., Jiang, Z. (2018). *J. Membr. Sci.*, **565**, 331–341.
15. Yang, H. Wu, Z. Yao, B. Shi, Z. Xu, X. Cheng, F. Pan, G. Liu, Z. Jiang, Cao, X. (2018). *J. Mater. Chem. A*, **6**, 583–591.
16. Peng, Y., Xu, G., Hu, Z., Cheng, Y., Chi, C., Yuan, D., Cheng, H., Zhao, D. (2016). *ACS Appl. Mater. Interfaces*, **8**, 18505–18512.
17. Wang, C., Li, Z., Chen, J., Li, Z., Yin, Y., Cao, L., Zhong, Y., Wu, H. (2017). *J. Membr. Sci.*, **523**, 273–281.

18. Li, C., Li, S., Tian, L., Zhang, J., Su, B., Hu, M. Z. (2019). *J. Membr. Sci.*, **572**, 520–531.

19. Kandambeth, S., Biswal, B. P., Chaudhari, H. D., Rout, K. C., Kunjattu, S. H., Mitra, S., Karak, S., Das, A., Mukherjee, R., Kharul, U. K. (2017). *Adv. Mater.*, **29**, 1603945.

20. Sasmal, H. S., Aiyappa, H. B., Bhange, S. N., Karak, S., Halder, A., Kurungot, S., Banerjee, R. (2018). *Angew. Chem. Int. Ed.*, **130**, 11060–11064.

21. Halder, A., Karak, S., Addicoat, M., Bera, S., Chakraborty, A., Kunjattu, S. H., Pachfule, P., Heine, T., Banerjee, R. (2018). *Angew. Chem. Int. Ed.*, **130**, 5899–5904.

22. Fan, H., Gu, J., Meng, H., Knebel, A., Caro, J. (2018). *Angew. Chem. Int. Ed.*, **57**, 4083–4087.

23. Fan, H., Mundstock, A., Feldhoff, A., Knebel, A., Gu, J., Meng, H., Caro, J. (2018). *J. Am. Chem. Soc.*, **140**, 10094–10098.

24. Fu, J., Das, S., Xing, G., Ben, T., Valtchev, V., Qiu, S. (2016). *J. Am. Chem. Soc.*, **138**, 7673–7680.

25. Abraham, J., Vasu, K. S., Williams, C. D., Gopinadhan, K., Su, Y., Cherian, C. T., Dix, J., Prestat, E., Haigh, S. J., Grigorieva, I. V. (2017). *Nat. Nanotechnol.*, **12**, 546.

26. Wang, H., Zeng, Z., Xu, P., Li, L., Zeng, G., Xiao, R., Tang, Z., Huang, D., Tang, L., Lai, C. (2019). *Chem. Soc. Rev.*, **48**, 488–516.

27. Li, G., Zhang, K., Tsuru, T. (2017). *ACS Appl. Mater. Interfaces*, **9**, 8433–8436.

28. Ying, Y., Liu, D., Ma, J., Tong, M., Zhang, W., Huang, H., Yang, Q., Zhong, C. (2016). *J. Mater. Chem. A*, **4**, 13444–13449.

29. Kuehl, V. A., Yin, J., Duong, P. H., Mastorovich, B., Newell, B., Li-Oakey, K. D., Parkinson, B. A., Hoberg, J. O. (2018). *J. Am. Chem. Soc.*, **140**, 18200–18207.

30. Dey, K; Pal, M., Rout, K. C., Kunjattu, S. H., Das, A., Mukherjee, R., Kharul, U. K., Banerjee, R. (2017). *J. Am. Chem. Soc.*, **139**, 13083–13091.

31. Matsumoto, M., Valentino, L., Stiehl, G. M., Balch, H. B., Corcos, A. R., Wang, F., Ralph, D. C., Marin˜as, B. J., Dichtel, W. R. (2018). *Chem*, **4**, 308–317.

32. Shinde, D. B., Sheng, G., Li, X., Ostwal, M., Emwas, A.-H., Huang, K.-W., Lai, Z. (2018). *J. Am. Chem. Soc.*, **140**, 14342–14349.

33. Kang, Z., Peng, Y., Qian, Y., Yuan, D., Addicoat, M. A., Heine, T., Hu, Z., Tee, L., Guo, Z., Zhao, D. (2016). *Chem. Mater.*, **28**, 1277–1285.

34. Yan, T., Lan, Y., Tong, M., Zhong, C. (2018). *ACS Sustainable Chem. Eng.*, **7**, 1220–1227.

35. Fan, H., Mundstock, A., Gu, J., Meng, H., Caro, J. (2018). *J. Mater. Chem. A*, **6**, 16849–16853.

36. Shan, M., Seoane, B., Rozhko, E., Dikhtiarenko, A., Clet, G., Kapteijn, F., Gascon, J. (2016). *Chem.-Eur. J.*, **22**, 14467–14470.

37. Tong, M., Yang, Q., Ma, Q., Liu, D., Zhong, C. (2016). *J. Mater. Chem. A*, **4**, 124–131.

38. Wu, X., Tian, Z., Wang, S., Peng, D., Yang, L., Wu, Y., Xin, Q., Wu, H., Jiang, Z. (2017). *J. Membr. Sci.*, **528**, 273–283.

Chapter 4

Heterogeneous Catalytic Application of COFs

Covalent organic frameworks (COFs) are defined as highly porous and crystalline polymers, constructed and connected via covalent bonds, extending along two or three dimensions. As compared with other porous materials, such as zeolite and active carbon, the versatile and alternative constituent elements, chemical bonding types, and characteristics of the ordered skeleton and pores of COFs enable the creation of more COFs for diverse applications, including gas separation and storage, optoelectronics, proton conduction, and energy storage and, in particular, "catalysis." The representative candidates of next-generation catalysis materials due to their large surface areas and accessible and size-tunable open nanopores, COFs are suitable for incorporating external useful active ingredients, such as ligands, complexes, and metal nanoparticles, and for substrate diffusion. This chapter will introduce and discuss examples of COFs for application in heterogeneous catalysts.

4.1 Heterogeneous Catalysts of COFs for C–C Bond Coupling Reactions

The catalytic C–C coupling method is the most important core of the current organic synthesis chemistry field. Synthetic projects (<85%)

Covalent Organic Frameworks
Atsushi Nagai

ISBN 978-981-4800-87-7 (Hardcover), 978-1-003-00469-1 (eBook)
www.jennystanford.com

in fine chemical production, pharmaceutical synthesis, agricultural chemicals, household chemicals, etc., are involved in catalyzed C–C coupling movements. Nowadays, catalytic C–C bond formation often resorts to homogeneous transition-metal complex catalysis systems such as the famous Pd-catalyzed Suzuki–Miyaura and Heck cross-coupling reactions [1]. Behind the success of these systems is the fact that these d-block transition-metal complexes commonly have empty π^* antibonding orbitals for π back-bonding and an electron pair for coordination to sp^2- or sp-hybridized carbon–carbon multiple bonds [2]. Moreover, the centers of these transition-metal complexes are adapted to oxidative addition and reductive elimination. The low-valent transition metals are stabilized by a variety of ligands bearing lone pairs of electrons and π^* antibonding orbitals. Thus, it is easy to choose a number of different chiral ligands to render the transition-metal complex asymmetric activity [3]. These chiral organometallic compounds are often very powerful catalysts for asymmetric C–C coupling reactions. The realization of this asymmetric cross-coupling has greatly enlarged the application scope of transition-metal-catalyzed reactions [4]. Although covalent organic frameworks (COFs) are new materials with a short history, the research on COFs is evolving rapidly. More and more research articles have revealed that COFs possess many unique properties, significantly differentiating them from organic-inorganic hybrids, such as metal-organic frameworks (MOFs) [5]. Why do chemists favor COFs specifically for a catalytic C–C coupling reaction? The reactions can be summarized as follows:

- From the perspective of industrial applications, a Pd-based homogeneous catalyst has an intrinsic limitation. Especially for drug synthesis, the biotoxicity of the residue from noble metals has long worried and been criticized by the public. And the efficiency of metal separation is unsatisfactory. Moreover, the cost of separation comprises a large portion of the total cost of drug production. Thus, developing heterogeneous catalysts, especially encapsulating catalytically active species such as Pd into the pore space of porous materials like COFs with uncompromising catalytic efficiency, will solve the issue of separating the toxic metal residues.

- From the perspective of catalytic efficiency, which is the core issue in catalysis, after binding metal active species, the metal catalytic center is confined by the COF's pores. On the one hand, this increases the difficulty of the substrate diffusing and approaching the catalytic sites and the products leaving the catalytic center, which would decrease the turn-over frequency (TOF). However, on the other hand, if the COF constituents and pore structures are fine-tuned, the COFs will repulse solvent molecules and accelerate substrate adsorption and product desorption. Compared with a homogeneous catalyst, this acceleration will greatly increase the efficiency and selectivity due to the pore effect. Certainly, realizing this point is challenging. However, this is the most attractive property of COFs, that is, to search for the optimal COF constituent and pore structure. On one hand, this optimized COF constituent and pore structure will settle and accommodate the metal catalytic center to exert its catalytic ability of oxidative addition and reductive elimination to the maximum. On the other hand, this will fine-tune the selectivity of substrate adsorption and product desorption rate. Most of the following examples in this chapter are the breakthroughs and proceedings in these aspects.
- Compared with other 2D or 3D pore structure materials, such as the most similar MOF counterparts, the advantage of COFs in catalyzing C–C coupling reaction is the absence of another external transition-metal center (metal in the MOF) to influence the encapsulated Pd catalysis, which avoids as-induced side reactions.
- As mentioned above, the stability to aqueous, acidic, basic, and organic solvents renders COFs inert to solvation and decomposition. Due to the short history of COFs, their applications and adaptions for C–C coupling reactions require further development and improvement. Even so, the currently reported examples, that is, the following recommended ones, have already displayed bright prospects for significant application and development.

The following examples and discussion are divided according to the C–C coupling reaction types and several typical COFs' catalyst synthesis and catalytic performance are recommended.

4.1.1 Suzuki–Miyaura Reaction

In 2011, Wang and coworkers demonstrated the first example of a COF for catalysis application [6]. An imine-linked 2D COF (COF-LZU1) was prepared by the condensation of 1,3,5-triformylbenzene and 1,4-diaminobenzene in a 1,4-dioxane/aqueous acetic acid solution in solvothermal conditions after liquid nitrogen flash freezing, evacuation, and flame-sealing treatment (Fig. 4.1). The COF-LZU1 took a 2D layered-sheet structure, with eclipsed nitrogen atoms in adjacent layers, at a distance of about 3.7 Å. This imine-linked COF-LZU1 demonstrated high coordination affinity to $Pd(OAc)_2$ due to strong nitrogen-palladium interaction. Only by simple impregnation, $Pd(OAc)_2$ was effectively incorporated into COF-LZU1 channels and pores. This palladium metal incorporation has no effect on the long-ordered structure of COF-LZU1. However, the intensity of powder X-ray diffraction (PXRD) and the Brunauer–Emmett–Teller (BET) surface area were reduced to a certain degree. The Pd-incorporated COF displayed very enhanced catalytic activity, a shorter reaction time, and a lower catalyst load than $Pd(OAc)_2$ and its Pd COF analogues in a typical Suzuki–Miyaura cross-coupling reaction (Fig. 4.2). Moreover, this COF did not lose its catalytic activity at all even after four cycles of reuse. This superior catalytic activity and stability after reutilization rendered it a very promising catalyst for a classical Suzuki–Miyaura reaction.

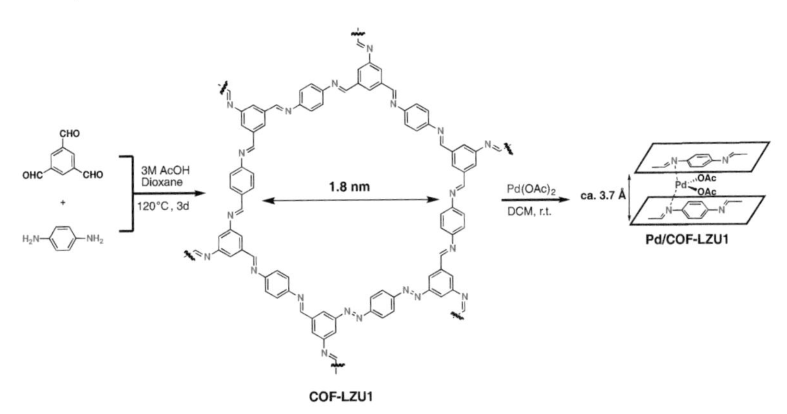

Figure 4.1 Schematic representation of the synthesis of COF-LZU1 and Pd/COF-LZU1.

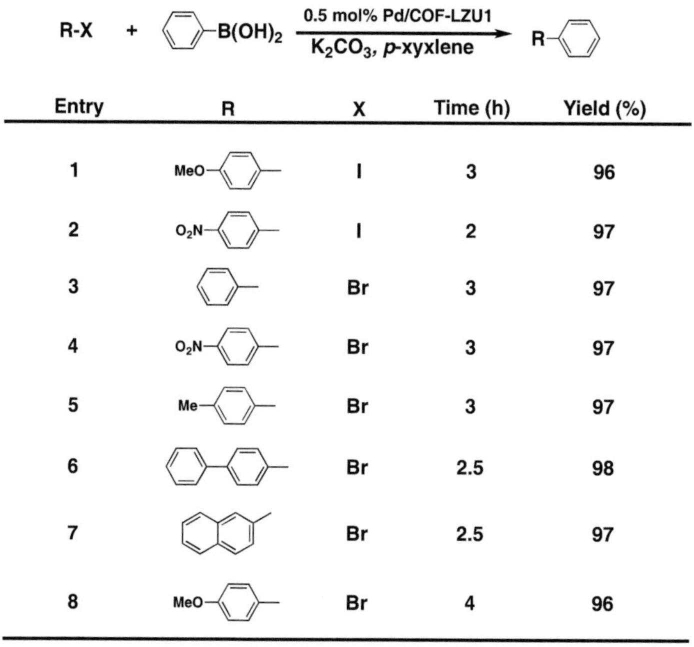

R-X + —$B(OH)_2$ —(0.5 mol% Pd/COF-LZU1; K_2CO_3, *p*-xyxlene)→ R—

Entry	R	X	Time (h)	Yield (%)
1	MeO—	I	3	96
2	O_2N—	I	2	97
3		Br	3	97
4	O_2N—	Br	3	97
5	Me—	Br	3	97
6		Br	2.5	98
7		Br	2.5	97
8	MeO—	Br	4	96

Figure 4.2 Catalytic activity test of Pd/COF-LZU1 in the Suzuki–Miyaura coupling reaction.

After the report by Wang and group, Jiang et al. also reported that a porphyrin-based H_2P-Bph COF could incorporate $Pd(OAc)_2$ species to efficiently catalyze a Suzuki reaction, with excellent yields, ranging from 97.1% to 98.5% [7]. This COF was prepared via condensation between 5,10,15,20-tetra(*p*-amino-phenyl)porphyrin and 4,4′-biphenyldialdehyde in an EtOH/mesitylene/acetic acid aqueous solution at 120°C for 3 days in vacuum (Fig. 4.3). This COF was nitrogen-rich due to the porphyrin unit's tetrapyrrole group and imine C=N bonds. These excess nitrogen groups aced as effective docking sites for $Pd(OAc)_2$ complexation. Solid state–^{13}C–nuclear magnetic resonance (NMR), Fourier-transform infrared spectroscopy, X-ray photoelectron spectroscopy, and inductively coupling plasma–atomic emission spectroscopy characterizations all confirmed the inclusion of $Pd(OAc)_2$. The porosity and crystallinity decreased to a certain degree after Pd incorporation. This Pd-H_2P COF showed superior catalytic activity for Suzuki cross-coupling reactions for a variety of bromoarenes and phenylboronic acid, forming biphenyl derivatives with yields ranging from 97.1% to 98.5%, surpassing the yields of its Pd MOF and Pd/C counterparts (the yield of 4-methosybiphenyl product was 65.0%, while the yield for Pd/MOF was 84.1% [8]). This report was inspiring and encouraging since the authors showed that a porphyrin-based catalyst could not only promote radical or carbene-based oxidation but also accelerate the cross-coupling reaction by incorporating a Pd complex into a COF support.

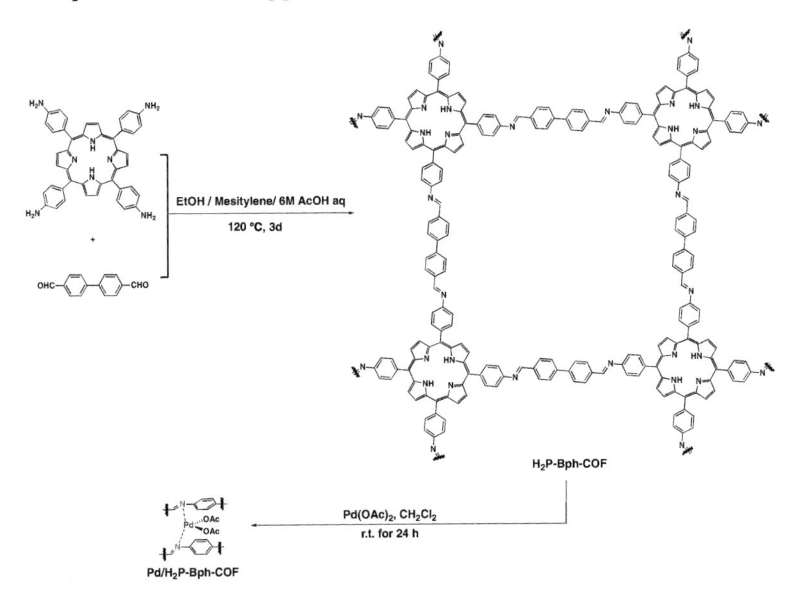

Figure 4.3 The development of H_2P-Bph COF and Pd/H_2-Bph COF.

Apart from imine and porphyrin COFs, a triazine-based COF could also serve as an efficient amphiphilic support for Pd(0) nanoparticle (NP) deposition [9]. The triazine COF was synthesized by the condensation reaction of 4,4′,4″-(1,3,5-triazine-2,4,6-triyl)tris(oxy) tribenzenealdehyde and benzene-1,4-diamine in 1,4-dioxane/ mesitylene in the presence of aqueous acetic acid as a catalyst by heating at 120°C for 72 h. The as-prepared triazine COF contained both a long and flexible appendage and a nitrogen- and oxygen-rich skeleton. The nitrogen-rich moiety was responsible for the facile in situ reduction of Pd^{2+} to Pd(0) without any external oxidants. The ether and imine moieties had strong interaction with Pd(0) NPs by stabilizing and dispersing them on the COF. This triazine COF was very effective in catalyzing the multifold Heck and Suzuki–Miyaura cross-coupling reactions. Unprecedentedly high turn-over numbers (TONs) and TOFs of the catalyst for multifold Heck reactions were provided as compared with its homogeneous Pd, Pd MOF, and Pd/C counterparts. Extremely short reaction times (1.5 h) and an excellent isolated yield (up to 99%) was observed. The recyclability of the catalyst showed apparently no loss in activity. This resulted in a viable strategy to dock noble metal NPs into a COF without external reducing agents, applying the triazine monomer itself as a reductant for efficient Heck and Suzuki–Miyaura cross-coupling reactions by appending long and flexible groups, leading to the formation of an amphiphilic structure and incorporating more nitrogen and oxygen atoms.

Two different kinds of Schiff-base COFs as triazine-based imine and β-ketoenamine-linked COFs (TAT-DHBD and TAT-TFP) could be synthesized from 1,3,5-tris(4-aminophenyl)triazine (TAT) and 2,5-dihydroxybenezne-1,4-dicarboxaldehyde (DHBD) or 1,3,5-triformylphloroglucinol (TFP) under solvothermal conditions in a dioxane/mesitylene mixture (Fig. 4.4) [10]. Including $Pd(OAc)_2$ into the pore space and between the interlayer region of the 2D sheets, Pd-loaded TAT-DHBD and TAT-TFA COFs were prepared. Sequentially, the Pd-loaded COFs were synthesized by the $NaBH_4$ reduction of Pd(II) to Pd(0) NPs. Four Pd-loaded COFs illustrated very good catalytic activity in a Suzuki–Miyaura cross-coupling reaction between bromobenzene and phenylboronic acid. The best performance was obtained with Pd(0)-TAT-TFP COF catalyst, which provided excellent conversion and yield for electron-rich and electron-deficient bromobenzenes. The most appropriate substrate was 4-cyanobromobenzene with phenylboronic acid, which showed an almost quantitative formation of 4-cyanobisphenyl only after 4 h by this Pd(0)-TAT-TFA COF catalyst. As a result, these Pd COF catalysts all displayed very good stability and reusability without apparent leaching of Pd and loss of activity.

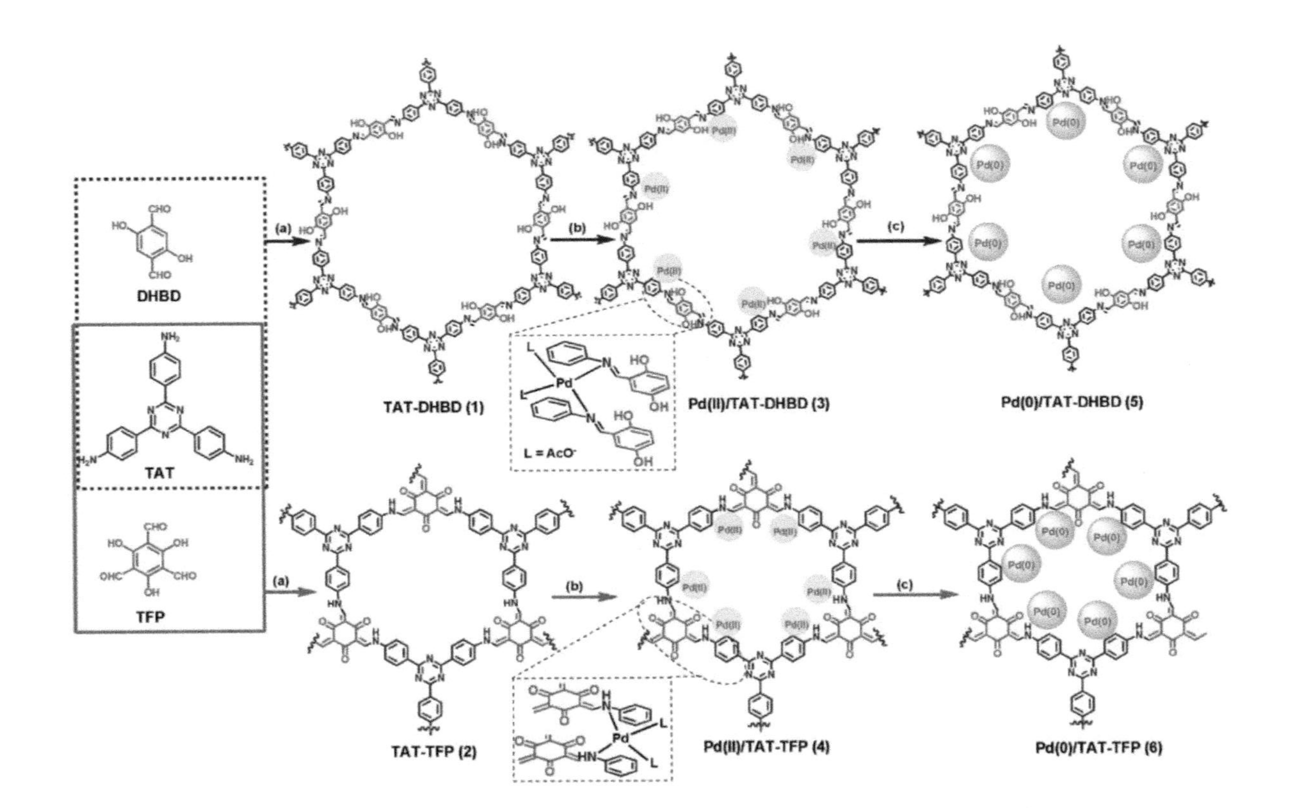

Figure 4.4 Synthesis of TAT-DHBD (1) and TAT-TFP (2) and their Pd-embedded COFs. Conditions: (a) dioxane/mesitylene, 6 M AcOH, 120°C, 3 days; (b) Pd(OAc)2, CH_2Cl_2, 24 h, RT; (c) $NaBH_4$, MeOH, 48 h, RT. Reproduced with permission from Ref. [10]. Copyright (2017), John Wiley and Sons.

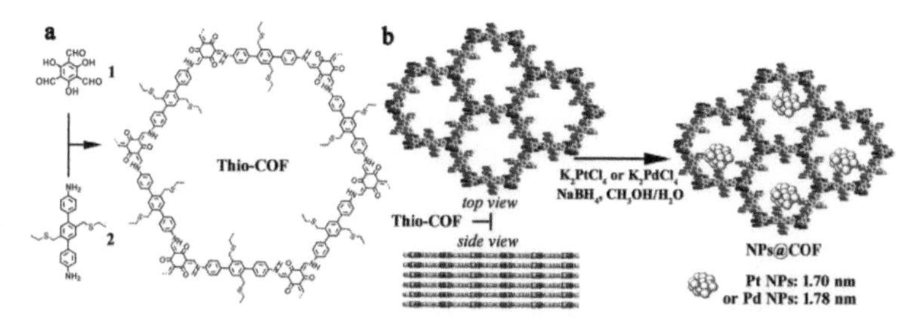

Figure 4.5 (a) Synthesis of thio-COF and (b) schematic representation of the synthesis of thio-COF-supported PtNPs@COF. Top and side views of the energy-minimized models of thio-COF (yellow, S; blue, N; gray, C; red, O) are shown in (b). Reprinted with permission from Ref. [11]. Copyright (2017) American Chemical Society.

More recently, a thioether-containing COF was reported to give excellent support to ultrafine Pt and Pd NPs, providing a very narrow size distribution and superior stability [11]. Inspired by their designed shape-persistent thioether-containing organic cage, which hosted ultrafine Pt and Pd NPs with a very narrow size distribution, Zhang and coworkers elaborately designed and prepared the PtNPs@COF and PdNPs@COF by condensing a trialdehyde and a thioether-containing diamine in dioxane/mesitylene/aqueous acetic acid solution at 120°C for 3 days (Fig. 4.5). The as-formed thio-COF was further complexed with K_2PtCl_4 and K_2PdCl_4 in an aqueous solution and then reduced by a methanolic $NaBH_4$ solution to PtNPs@COF and PdNPs@COF. According to a series of structural, morphological, and compositional characterization, the authors demonstrated that ultrafine Pd and Pt NPs were uniformly incorporated into the pore space of the thio-COF. The thioester functional group provided strong metal-sulfur interaction to stabilize the ultrafine noble metal NPs to prevent them from aggregation. Moreover, the long-range ordered pore-channel structure also assisted in the stabilization of the residing external NPs. The as-formed PtNPs@COF and PdNPs@COF illustrated very good catalytic activity toward 4-nitrophenol reduction and Suzuki–Miyaura cross-coupling between a variety of aryl halides and phenylboronic acid. PdNPs@COF provided excellent NMR yields (up to 99%) for the cross-coupling of 4-methyl-iodobenzene and phenylboronic acid to form 4-methylbiphenyl. Furthermore, these two COF-based catalysts presented excellent stability and recyclability after simple centrifugation or natural settling for the cycling test, almost no decrease in conversion and yield in catalytic performance after the fifth cycle test, and no noticeable leaching of metal NPs. This example is significant because it was the first use of a thioether functional group to stabilize the narrowly distributed ultrafine noble metal NPs for effective cross-coupling reactions in a COF support.

4.1.2 Heck, Sonogashira, and Silane-Based Cross-Coupling Reactions

A Heck cross-coupling reaction is the Pd-catalyzed coupling of aromatic or vinylic halides with an unsaturated olefinic C=C bond, as shown in Scheme 4.1 [12]. The catalytic cycle comprises oxidative addition of aromatic halides, coordination with an olefinic C=C bond, *cis*-insertion, *cis*-β-hydride elimination, and reductive elimination. This reaction requires a Pd(0) active species. The advantage of a Heck reaction is its high regio- and stereoselectivity. In contrast, the disadvantage is the costly Pd catalyst. The Sonogashira coupling reaction is the Pd/Cu cocatalyzed coupling of a terminal alkyne with aryl or vinylic halides to a Pd(0) center. Then, a Cu-amine complex mediates the transmetalation reaction. The last step is a reductive elimination, releasing the coupling product and regenerating the Pd(0) catalyst. A Cu-amine complex acts as a cocatalyst to assist in deprotonating the alkyne substrate.

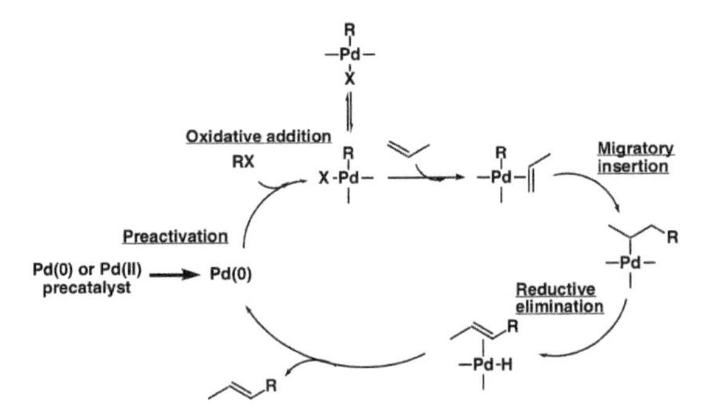

Scheme 4.1 Pd-catalyzed Heck reaction and its catalytic cycle [12].

Banerjee and coworkers demonstrated that the introduction of a large number of nitrogen and oxygen atoms into the skeleton of a COF would reinforce its stability when metal NPs or complexes are deposited. They exhibited that the condensation reaction of TFP and paraphenylenediamine in mixed solvents (mesitylene/dioxane) in the presence of aqueous acetic acid at 120°C for 3 days in an inert atmosphere would generate a nitrogen- and oxygen-rich COF (Fig. 4.6) [13]. The as-synthesized imine COF is further deposited with Pd(II) complexes by immersing it in a methanol solution containing $Pd(OAc)_2$. Pd(0) NPs are generated from the in situ reduction of Pd(II) COF with $NaBH_4$. The authors demonstrated that Pd(0) and Pd(II) inclusion does not greatly change the crystallinity and flower-like morphology of the COF. Further, the Pd(0) COF shows superior catalytic activity toward Heck- and Sonogashira-type reactions. The Pd(II) COF demonstrated a considerably robust catalytic ability for intramolecular C–H activation and further C–C coupling reaction, synthesizing 9H-carbazole from diphenylaniline. This report manifested its significance in incorporating Pd(0) and Pd(II) into the same COF and applied the metal–COF composite in highly selective C–C coupling and C–H activation transformations.

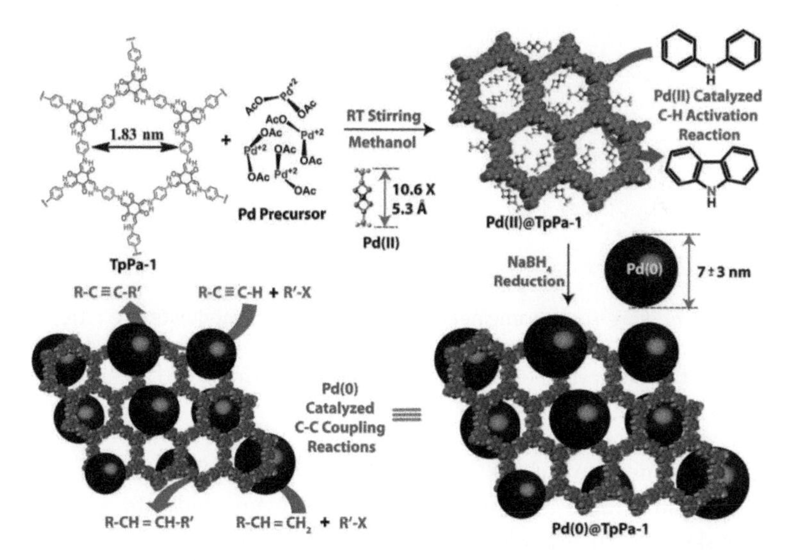

Figure 4.6 Development of Pd(II)- and Pd(0)-doped COFs (i.e., Pd(II)@TpPa-1 and Pd(0)@TpPa-1) and summary of their catalytic activity toward Sonogashira, Heck, and oxidative biaryl couplings. The doped Pd(0) nanoparticles are probably situated on the TpPa-1 surface. The scheme is to represent the synthesis and organization of Pd nanoparticles on COF (TpPa-1) and it is not exactly fit to scale. Republished with permission of Royal Society of Chemistry, from Ref. [13], copyright (2014); permission conveyed through Copyright Clearance Center, Inc.

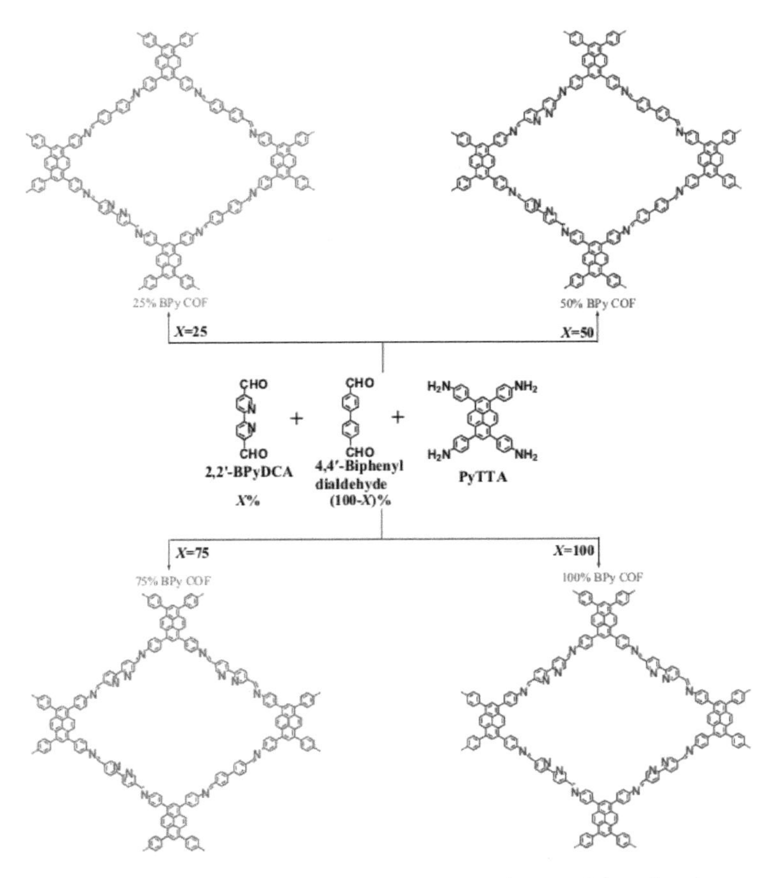

Figure 4.7 Pore surface engineering strategy used to modulate the nitrogen content of the 2D imine–linked COFs and scheme for regulated $Pd(OAc)_2$ coordination on bipyridine and imine groups. Reprinted from Ref. [14], Copyright (2016), with permission from Elsevier.

Chai and coworkers reported that two different nitrogen ligands bipyridine and imine could be incorporated into a single COF skeleton to provide differentiated Pd coordinating sites [14]. The authors designed and prepared *X*% bpy COF by condensing *x*% 2,2-bipyridine-5,5′-dicarbaldehyde, 100-*X*% 4,4′-biphenyl dialdehyde, and 4,4′,4″,4‴-(pyrene-1,3,6,8-tetrayl)tetraaniline (PyTTA) building blocks in mesitylene/dioxane mixed solvents in the presence of acetic acid (3 M) at 120°C for 3 days, as shown in Fig. 4.7. The as-formed *X*% bpy COF contained two different kinds of nitrogen ligands, bipyridine and imine. With further $Pd(OAc)_2$

complexation experiment results, they demonstrated that $Pd(OAc)_2$ coordinated with both bipyridine and imine units, but in different regions. $Pd(OAc)_2$ combined with bipyridine mainly dwelled in the pore space, while $Pd(OAc)_2$ joined with imine resided between the adjacent layers of the 2D COF. Furthermore, the authors demonstrated that these Pd@bpy COFs displayed a very good catalytic performance toward the classical Pd-catalyzed Heck reaction between a series of aryl halides and styrene. The Pd(II)@75% bpy COF showed the best catalytic ability, providing >90% yield after four consecutive runs. The superior activity for the Heck reaction of these Pd@bpy COF catalysts was attributed to the uniform dispersion and the ultrahigh loading of $Pd(OAc)_2$.

On other occasions, Pd(0) NPs could be generated in situ by choosing a predesigned metal-anchored building block. Initially, a 2,2-bipyridine-5,5′-diamine palladium chloride (Bpy-$PdCl_2$) complex was formed by a condensation reaction between $PdCl_2$ and 2,2′-bipyridine-5,5′-diamine (Fig. 4.8) [15]. Then, via Schiff-base condensation between Bpy-$PdCl_2$ and Tp, a Pd@TpBpy COF was prepared. This in situ generated Pd@TpBpy COF did not require any external reducing agents for Pd(II) reduction to Pd(0) NPs. The as-formed Pd@TpBpy COF represented excellent catalytic performance toward a tandem C–C and C–O bond formation reaction between 2-bromophenol and phenylacetylene. The Pd@TpBpy COF heterogenous catalyst promoted tandem cyclization and provided various substituents on the phenylacetylene or 2-bromophenol. Moreover, this ketoenamine and bipyridine anchored Pd@TpBpy COF showed very good stability and recyclability.

A triazine-based COF-SUD1-palladium hybrid could be an active catalyst for the cross-coupling between silanes and aryliodides [16]. The COF-SDU1 was prepared via an imine condensation from tri-(4-formacylphenoxy)-1,3,5-triazine and *p*-phenylenediamine in *o*-dichlorobenzene/*n*-butanol/6 M AcOH with heating at 85°C for 7 days. The as-obtained COF contained two kinds nitrogen ligand azine and imine, which are both suitable coordinating sites for Pd(II). After a simple solution infiltration, monodispersing Pd(II) ion was docked in the 2D COF. This Pd(II)-COF-SDU1 showed excellent catalytic activity for a one-pot silane oxidation reaction. The transformation of phenylsilanes with aryliodides to biphenyls was effectively catalyzed in a methanolic solution. A variety of electron-rich and electron-deficient aryliodides were cross-coupled with phenylsilanes and gave good to excellent yields. This COF also displayed good recyclability and reusability without detectable Pd leaching and loss of activity.

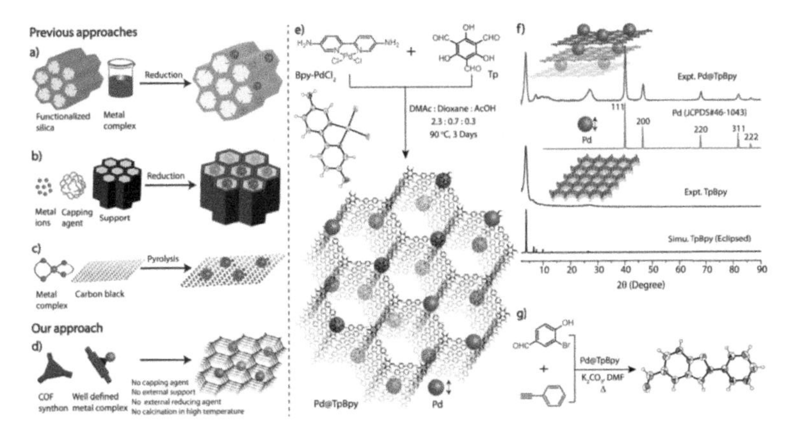

Figure 4.8 Various strategies are used to convert a homogeneous catalyst into a reversible heterogeneous version (a–c) Reported approaches and (d) new approach. (e) Synthesis details of the in situ generation of highly dispersed Pd nanoparticles in the TpBpy skeleton. The size of the Pd nanoparticles and the COF pore organization are not exactly to scale. (f) Comparison of PXRD patterns of simulated TpBpy (black) with experimental TpBpy (blue), Pd nanoparticles (cyan), and experimental Pd@TpBpy (red). (g) Schematic representation of tandem catalysis by Pd@TpBpy. Reprinted with permission from Ref. [15]. Copyright (2017) American Chemical Society.

The important point of COFs' successful application in Heck and Sonogashira cross-couplings are summarized as follows: low Pd loading, high catalytic efficiency, and facile desorption of halide ions.

4.2 Chiral Heterogeneous Catalysts of COFs for Asymmetric C–C Bond Coupling Reactions

In 2014, Jiang et al. developed a chiral-organocatalytic COF prepared through pore surface engineering, whose system was proposed in 2011 [17, 18]. This organocatalytic COF was prepared by integrating a chiral pyrrolidine unit into the main chain of the porphyrin-imine COF, as shown in Fig. 4.9. Initially, they introduced an ethynyl group into the imine moiety. By the facile alkyne-azide click reaction, they anchored the triazole-substituted pyrrolidine ring to the imine by a post-treatment catalyzed by CuI. The as-prepared COF demonstrated its activity for organocatalysis and displayed a variety of advantages in catalyzing an enantioselective asymmetric Michael addition reaction. The conversion of the reactants was much accelerated by this organocatalytic COF; however, moderate diastereoselectivity (d.r.) and enantiomeric excess (e.e.) values were obtained (Fig. 4.10). The most important was that this was the first time COFs realized enantioselectivity control in a catalytic organic synthetic reaction.

Moreover, the design and synthesis of another organocatalytic COF was also reported by Jiang et al. [19]. This was a milestone event for a COF applied for an organic synthetic purpose. They discovered that the introduction of a methoxy group in the edge unit would greatly increase the stability of the COF against humidity, acidity, and basicity since the methoxy group reinforced the interlayer interaction. The COF was prepared by condensing triphenylbenzenetriamine (TPB), 2,5-*bis*(2-propynyloxy)terephthalaldehyde (BPTA) and 2,5-dimenthoxyterephthaladehyde (DMTA) (Fig. 4.11). The $[HC{\equiv}C]_x$-TPB-DMTA COFs with alkynyl groups were further transformed to $[(S)\text{-Py}]_x$-TPB-DMTP COFs by a postsynthetic click reaction with (*S*)-2-(azideomethyl)pyronylidine catalyzed by a CuI catalyst. The as-formed $[(S)\text{-Py}]_x$-TPB-DMTA COFs demonstrated extremely strong stability when soaked in boiling water, 12 M HCl, and a 14 M NaOH solution. The COF displayed very little reduction in crystallinity and porosity after these harsh condition treatments. The intensities of the X-ray diffraction peaks showed no apparent decrease and the BET and Langmuir surface area almost remained unchanged. Besides extraordinary stability and uncompromised crystallinity and porosity, the most significant point of this COF was its functionality to catalyze chiral asymmetric organic reactions. For a typical organocatalytic asymmetric Michael addition between unactivated cyclohexanones and nitrostyrenes, this COF showed superior catalytic activity compared with its homogeneous counterpart (Fig. 4.12). The COF required only half the time required by the organocatalyst. After a five-cycle reusability test, this COF did not show any apparent loss in catalytic activity. The shining point of this COF was its combination of unprecedented stability, good crystallinity, and highly developed mesoporous structure with very powerful catalytic ability, accelerated reaction kinetics, and excellent yields and d.r. and e.e. values for chiral asymmetric Michael addition in an aqueous solution (Fig. 4.11d).

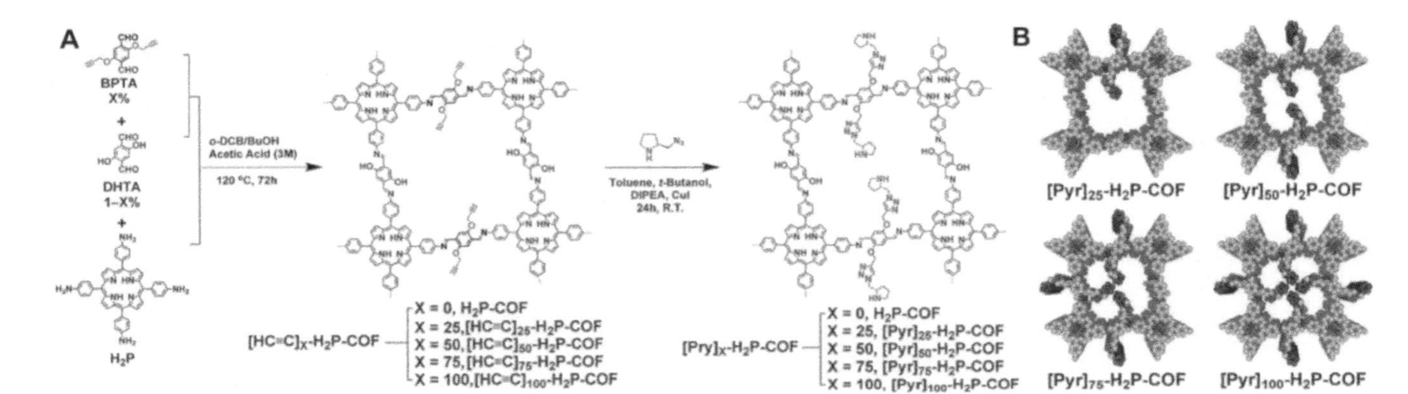

Figure 4.9 (A) The general strategy for the pore surface engineering of imine-linked COFs via a condensation reaction and click chemistries (the case for $X = 50$ was exemplified). (B) A graphical representation of $[Pyr]_x$-H_2P COF with catalytic sites of different densities on the pore walls (gray: carbon; red: nitrogen; green: oxygen; purple: carbon atoms of the pyrrolidine units; hydrogen is omitted for clarity). Republished with permission of Royal Society of Chemistry, from Ref. [18], copyright (2014); permission conveyed through Copyright Clearance Center, Inc.

	Time for 100% conversion (h)	dr	ee (%)
Control	3.3	60/40	49
$[Pyr]_{25}$-H_2P-COF	1	70/30	49
$[Pyr]_{50}$-H_2P-COF	2.5	70/30	50
$[Pyr]_{75}$-H_2P-COF	5	70/30	51
$[Pyr]_{100}$-H_2P-COF	9	65/35	44
Amorphous polymer 1	43	70/30	48
Amorphous polymer 2	65	65/35	46

Figure 4.10 Comparison of the pyrrolidine control, amorphous nonporous polymers, and COFs as catalysts for a Michael addition reaction. Republished with permission of Royal Society of Chemistry, from Ref. [18], copyright (2014); permission conveyed through Copyright Clearance Center, Inc.

Apart from above-mentioned chiral organocatalyst-incorporated COF for asymmetric organic transformations and noble metal–incorporated COFs for achiral organic transformations, a homochiral organocatalytic COF skeleton could also be a perfect support for organometallic Pd species with chiral ligands [20]. The Pd NPs dispersed in a chiral COF (CCOF) skeleton could effectively be a heterogeneous catalyst for asymmetric Henry and reductive Heck reactions, providing excellent isolated yields and e.e. values. The CCOF was prepared by a condensing reaction between cyanuric chloride and *S*-(+)-2-methylpiperazine with K_2CO_3 in a dioxane solution. Further, Pd(0) NPs were included in the CCOF by in situ reduction of a $Pd(NO_3)_2$ methanolic solution with $NaBH_4$ in a CCOF aqueous suspension (Fig. 4.13). The as-formed Pd@CCOF was uniformly dispersed between the CCOF 2D layer, not residing in the pore space due to the large size of the Pd NPs (2–5 nm) compared with the CCOF micropore (1.5 nm). The incorporation of Pd NPs greatly influenced the porous structure of the CCOF, enlarging its BET surface area and pore size. Furthermore, the authors demonstrated the synergistic catalytic activity by subjecting the Pd@CCOF catalyst in typical Henry and reductive Heck reactions. To the authors' delight, the Pd@CCOF catalyst displayed extremely superior catalytic ability toward these two reactions, providing an excellent yield (up to 99%) and a perfect e.e. value (up to 97%). To thc best of my knowledge, this was the first example of a heterogeneous

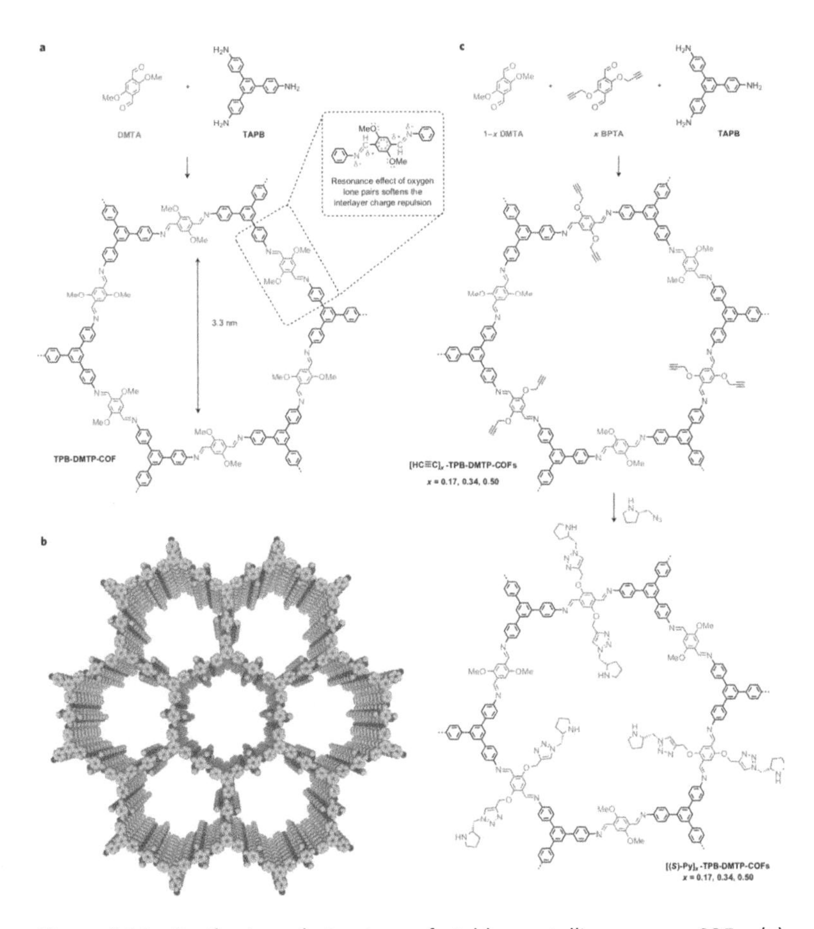

Figure 4.11 Synthesis and structure of stable crystalline porous COFs. (a) Synthesis of TPB-DMTP COF through the condensation of DMTA (blue) and TAPB (black). Inset: The structure of the edge units of TPB-DMTP COF and the resonance effect of the oxygen lone pairs that weaken the polarization of the C=N bonds and soften the interlayer repulsion in the COF. (b) Graphic view of TPB-DMTP COF (red, O; blue, N; gray, C; hydrogen is omitted for clarity). (c) Synthesis of chiral COFs ([(S)-Py]*x*-TPB-DMTP COFs, $x = 0.17, 0.34$, and 0.50; blue, DMTA; black, TAPB; red, BPTA; green, (S)-Py sites) via channel-wall engineering using a three-component condensation followed by a click reaction. Reprinted by permission from Springer Nature Customer Service Centre GmbH: Springer Nature, *Nature Chemistry*, Ref. [19], copyright (2015).

Pd-catalyzed chiral asymmetric reductive Heck reaction. The Pd@CCOF illustrated satisfactory recyclability and reusability after five cycles of reuse, without apparent loss of catalytic activity, providing an isolated yield of up to 93% and an e.e. value of 91% for the fifth cycle test. This development of Pd@CCOF manifested its importance in that it was the first COF combining a noble metal catalyst with a chiral organocatalyst in a single COF carrier to fulfill asymmetric transformations previously catalyzed by homogeneous organometallic compounds with complex and elaborately designed chiral ligands.

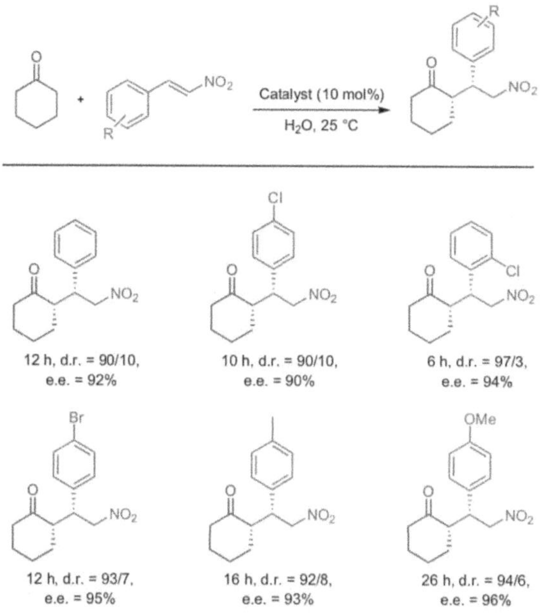

Catalyst (10 mol%)
H_2O, 25 °C

12 h, d.r. = 90/10, e.e. = 92%

10 h, d.r. = 90/10, e.e. = 90%

6 h, d.r. = 97/3, e.e. = 94%

12 h, d.r. = 93/7, e.e. = 95%

16 h, d.r. = 92/8, e.e. = 93%

26 h, d.r. = 94/6, e.e. = 96%

Figure 4.12 Scope of reactants. Different β-nitrostyrene derivatives investigated for the Michael reactions catalyzed with chiral COFs, their products, e.e. yields and d.r. values (red, cyclohexanone; green, newly formed C–C bond; blue, nitrostyrene derivatives). R, substituent H, Cl, Br, Me, or OMe. Reprinted by permission from Springer Nature Customer Service Centre GmbH: Springer Nature, *Nature Chemistry*, Ref. [19], copyright (2015).

More recently, Cui and coworkers reported a multivariate strategy to prepare CCOFs with controlled crystallinity and stability for asymmetric catalysis [21], in which crystallizing mixtures of triamines with and without chiral organocatalysts and a dialdehyde produced a series of two- and three-component 2D COFs (Fig. 4.14). These COFs were found to be efficient heterogeneous catalysts in

a)

Run	Ar-CHO	Yield (%)	*ee* (%)	Config.
1	Ph-CHO	97	95	R
2	4-(Me)-Ph-CHO	88	81	R
3	3-(Me)-Ph-CHO	87	82	R
4	4-(MeO)-Ph-CHO	82	81	R
5	3-(MeO)-Ph-CHO	80	79	R
6	4-(NO_2)-Ph-CHO	99	97	R
7	3-(NO_2)-Ph-CHO	94	88	R
8	4-(F)-Ph-CHO	96	95	R
9	4-(Br)-Ph-CHO	92	93	R
10	2-(Br)-Ph-CHO	90	90	R
11	4-(Cl)-Ph-CHO	94	95	R
12	4-(CN)-Ph-CHO	99	97	R
13	4-(OH)-Ph-CHO	96	90	R
14	biphenyl-CHO	5	ND[c]	ND
15	anthracene-CHO	ND	ND	ND

b)

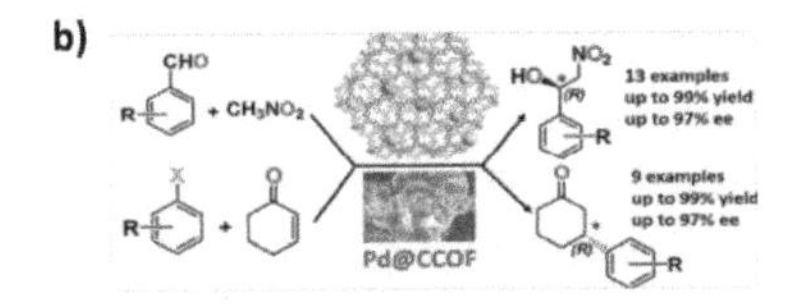

Figure 4.13 (a) Henry reactions of nitromethane with different aromatic aldehydes catalyzed by Pd@CCOF and (b) Pd NP-loaded homochiral COF for heterogeneous asymmetric catalysis. Reprinted with permission from Ref. [20]. Copyright (2017) American Chemical Society.

catalyzing the asymmetric amino-oxylation reaction, aldol reaction, and Diels–Alder reaction, with the stereoselectivity comparable to or surpassing their homogeneous analogues. Moreover, they developed a metal-directed synthesis strategy through which chiral Zn(salen)-based COFs can be obtained by the imine-condensations of enantiopure 1,2-diaminocyclohexane and *C*3-symmetric trisalicylaldehydes with one or no 3-*tert*-butyl group, as shown in Fig. 4.15. They found that the bulky tributyltrisalicylaldehydes containing CCOF possessed superior stability. These two COFs were further complexed with a variety of metal ions, such as Zn, Fe, Mn, Cr, V, and Co. The Mn(salen) modules in the COFs, Mn-salen COFs, demonstrated very good catalytic activity to a variety of chiral asymmetric organic synthetic reactions, such as V-salen COF–catalyzed cyanation reaction of aldehydes, Diels–Alder cycloaddition reaction catalyzed by a Co-salen COF, epoxidation catalyzed by Fe-salen and Mn-salen COFs, aminolysis opening of epoxides catalyzed by a Cr-salen COF, and the tandem one-pot heterogenous catalyst. These M(salen) CCOF–catalyzed reactions not only provided satisfactory yields but also realized excellent controls of enantioselectivity and diastereoselectivity. Moreover, good recyclability and reusability was proved in the case of V-salen COF–catalyzed cyanation of aldehydes, which showed almost no loss of enantioselectivity and conversion after five runs of cycle tests.

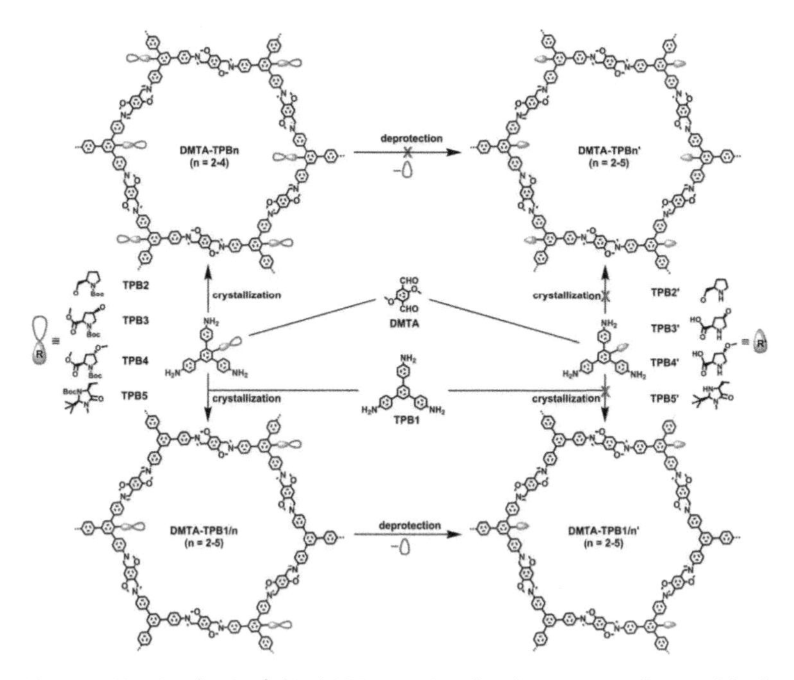

Figure 4.14 Synthesis of chiral COFs. Reprinted with permission from Ref. [21]. Copyright (2017) American Chemical Society.

The construction of functional COFs by a bottom-up strategy is a relatively difficult task because it must simultaneously meet the requirements for crystallinity and functionality. Even so, several attempts have been shown to be successful for the synthesis of COFs bearing catalytic sites with this method. For example, a sulfonated building lock, 2,5-diaminobenzenesulfonic acid (DABA), was used to construct a sulfonated COF together with another building block, TFP [23]. The as-synthesized TFP-DABA was found to be a highly efficient acid catalyst for fructose conversion with remarkable yields (97% for 5-hydroxymethylfurfural and 65% for 2,5-diformylfuran), good chemoselectivity, and good recyclability. Cui and coworkers reported the synthesis of two 2D CCOFs by direct imine condensations of enantiopure TADDOL-derived tetraaldehydes with 4,4′-diaminophenylmethane, which, after treatment with $Ti(OiPr)_4$ show highly catalytic activity for the asymmetric addition of diethylzinc to aldehydes [24].

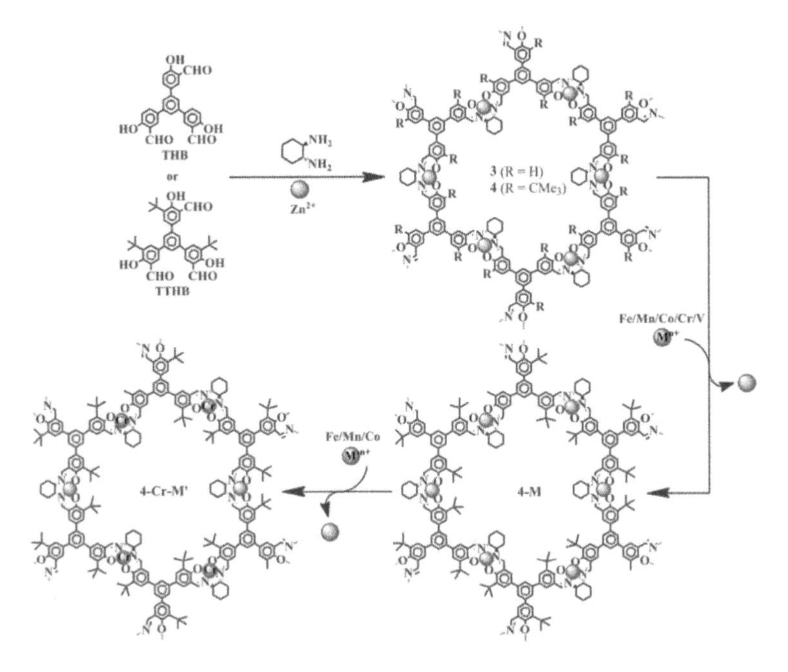

Figure 4.15 Synthesis of chiral COFs. Reprinted with permission from Ref. [22]. Copyright (2017) American Chemical Society.

In a similar way, Wang et al. reported a facile strategy for the direct construction of chiral-functionalized COFs using chiral (*S*)-4,4′-(2-(pyrrolidine-2-yl)-1H-benzimidazole as a building block (Fig. 4.16a) [25]. Two CCOFs, LZU-72 and LZU-76, are obtained based on the building block, as shown in Fig. 4.16. The catalytic activity of chemically stable LZU-76 was evaluated in the asymmetric aldol reaction, and the reaction afforded the desired aldol products with excellent enantioselectivity (88.4:11.6-94.0:60 e.r.), comparable to the enantioselectivity (93.1:6.9) of a model catalyst, (*S*)-4,7-diphenyl-2-(pyrrolidin-2-yl)-1H-benzimidazole (Fig. 4.16b).

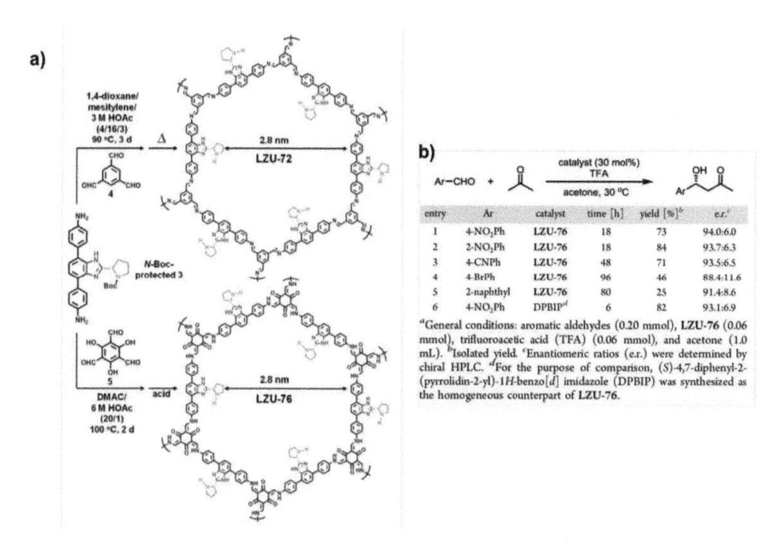

entry	Ar	catalyst	time [h]	yield [%][b]	e.r.[c]
1	4-NO_2Ph	LZU-76	18	73	94.0:6.0
2	2-NO_2Ph	LZU-76	18	84	93.7:6.3
3	4-CNPh	LZU-76	48	71	93.5:6.5
4	4-BrPh	LZU-76	96	46	88.4:11.6
5	2-naphthyl	LZU-76	80	25	91.4:8.6
6	4-NO_2Ph	DPBIP[d]	6	82	93.1:6.9

[a]General conditions: aromatic aldehydes (0.20 mmol), **LZU-76** (0.06 mmol), trifluoroacetic acid (TFA) (0.06 mmol), and acetone (1.0 mL). [b]Isolated yield. [c]Enantiomeric ratios (e.r.) were determined by chiral HPLC. [d]For the purpose of comparison, (*S*)-4,7-diphenyl-2-(pyrrolidin-2-yl)-1*H*-benzo[*d*] imidazole (DPBIP) was synthesized as the homogeneous counterpart of **LZU-76**.

Figure 4.16 (a) Schematic of the direct formation of chiral LZU-72 and LZU-76 and (b) catalytic activity check of LZU-76 in an asymmetric aldol reaction. Reprinted with permission from Ref. [25]. Copyright (2016) American Chemical Society.

4.3 Heterogeneous Bimetallic or Bifunctional Catalysts of COFs

A bifunctional organocatalytic COF was initially designed and realized by Banerjee et al., which was stable in aqueous and acidic conditions [26]. This COF was synthesized through the Schiff-base condensation between 2,3-dihydroxyterephthalaldehyde (2,3-Dha) and 5,10,15,20-tetrakis(4-aminophenyl)-21*H*,23*H*-porphyrin unit (Tph) in the mixed solvents of dichlorobenzene (*o*-DCB) and dimethylacetamide in the presence of 6 M acetic acid as a catalyst. The as-synthesized COF showed unprecedented stability in water and an acidic solution due to the catechol group, the presence of *trans* conformation of imine bonds, and intermolecular hydrogen bonding (–OH•N=C; D = 2.579, d = 1.858 Å, y = 146.11), which have been confirmed from the monomer crystal structure (Fig. 4.17).

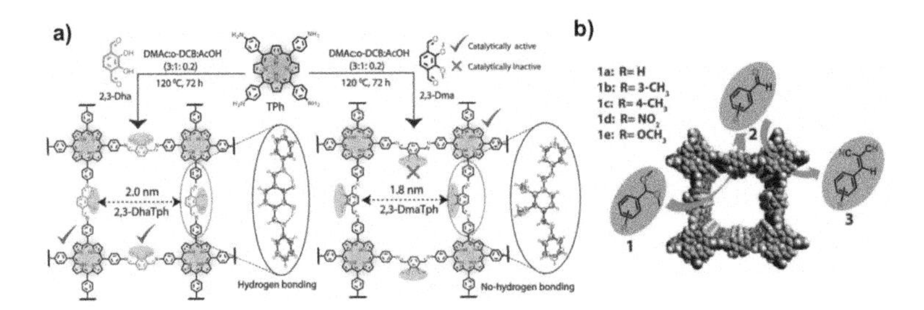

Figure 4.17 (a) The synthesis of 2,3-DhaTph and 2,3-DmaTph by the condensation of Tph and 2,3-Dha/2,3-Dma. The catalytically active porphyrin and catecholic –OH groups are shown in coral and cyan colors, respectively. An ORETP diagram of 2,3-DhaTph and 2,3-DmaTph monomer units. (b) The catalytic activity toward acid-base catalyzed reaction with various reactions. Republished with permission of Royal Society of Chemistry, from Ref. [26], copyright (2015); permission conveyed through Copyright Clearance Center, Inc.

Moreover, since the COF was considered from the Dha unit with weak acidic catechol groups and Tph group containing basic pyrrole groups and imine C=N bonds, this COF possessed acidic and basic sites, providing it a promising bifunctional heterogeneous catalyst. In a model cascade deacetalization–Knoevenagel reaction, this COF demonstrated an excellent isolated yield up to 96%. The deacetalization of benzaldehydedimethylacetal was catalyzed by the acidic sites of this COF catalyst, while a further Knoevenagel reaction between benzaldehyde and amlonitrile was effectively accelerated by the basic sites of the DhaTph COFs. This discovery manifested its significance in that it was the first stable bifunctional COF catalyst in water and an acidic solution.

With the exception of single metal–deposited COFs, bimetallic docked COFs were designed, synthesized, and applied as effective catalysts for a Heck–epoxidation tandem reaction. Mn and Pd bimetallic docking to a bipyridine-imine COF could be realized by a programmed synthetic procedure (Fig. 4.18) [27]. First, a Py-2,2′-BPyPh COF skeleton was constructed via a Schiff-base condensation reaction between PyTTA and 2,2′-bipyridyl-5,5-dialdehyde. The as-formed COF contained two different types of organic transformations, that is, Pd-catalyzed Heck cross-coupling reaction and Mn-catalyzed epoxidation reaction. The COF transformed iodobenzene and styrene to trans-stilbene oxide in a tandem reaction. Initially, Pd(OAc) incorporated in COF transformed iodobenzene and styrene into trans-stilbene with an excellent yield (up to 95%), while the Mn in COF catalyzed the epoxidation reaction of trans-stilbene to trans-stilbene oxide in an almost quantitative yield (99%). The control group experiments proved that Mn@Py-2,2′-BPyPh COF and Pd@Py-2,2′-BPyPh COF alone could only catalyze the separate epoxidation and Heck reaction. This finding was important since it demonstrated that by choosing elaborately designed ligands, different metal species could be incorporated into a single COF skeleton to fulfill different genres of organic transformation by certain metals.

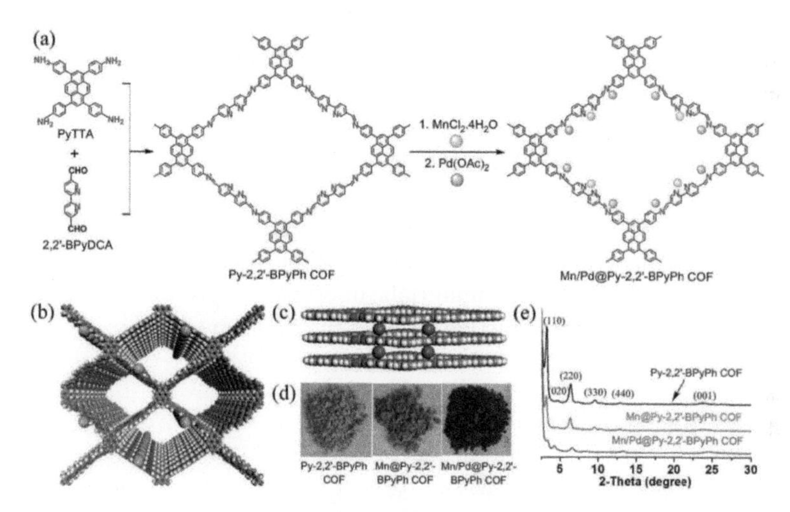

Figure 4.18 (a) Schematic representation of Mn/Pd bimetallic docked COFs prepared via a programmed synthetic procedure, (b) top view and (c) side view of Mn/Pd@Py-2,2′-BPyPh COF, and (d) appearances and (e) PXRD pattern of the COFs before and after metallic loading treatment. Republished with permission of RSC Publishing, from Ref. [27], copyright (2016); permission conveyed through Copyright Clearance Center, Inc.

Besides Mn and Pd codocking for a bimetallic COF catalyst, Rh and Pd bimetallic docking could also be realized via this 2D BPy COF (Fig. 4.19) [28]. By condensing PyTTA and 2,2′-bipyridy-5,5-dialdehyde and 100-X% 4,4′-biphenyldialdehyde in a three-component solvent, a series of structurally tunable 2D COFs were prepared. The authors demonstrated that by a further solution-infiltration method, Pd(OAc)2 and Ph(COD)Cl were sequentially incorporated into the COF skeleton in a programmed synthetic procedure. Pd(OAc)2 dispersed in the interlayer space coordinated with both imine units and bipyridine ligands, while the more structurally rigid Rh(COD)Cl was deposited in the pore space and complexed with bipyridine ligands. This Rh/Pd bimetallic–docked Bph COF demonstrated superior catalytic activity toward a tandem addition-oxidation reaction between phenylboronic acid and benzaldehyde to initially form the intermediate diphenylmethanol and further be oxidized to the final compound, benzophenone. The authors found that the Rh(COD)Cl moiety in the COF was accountable for the addition reaction between phenylboronic acid and benzaldehyde, surpassing its homogeneous Rh(COD)Cl analogue in catalytic activity. Pd(OAc)2 was responsible for the oxidation from diphenylmethanol to benzophenone. The as-prepared Rh/Pd-Bph COF showed excellent recyclability and reusability, providing isolated benzophenone products up to 85% yield even after five cycles of reuse, without noticeable leaching of the metal and apparent loss of activity. This report manifested its significance in that it illustrated that two different kinds of organometallic compounds could be docked in a structurally its tunable COF by different coordinating groups to render the COF materials different catalytic ability toward totally differentiated reaction types in the first time.

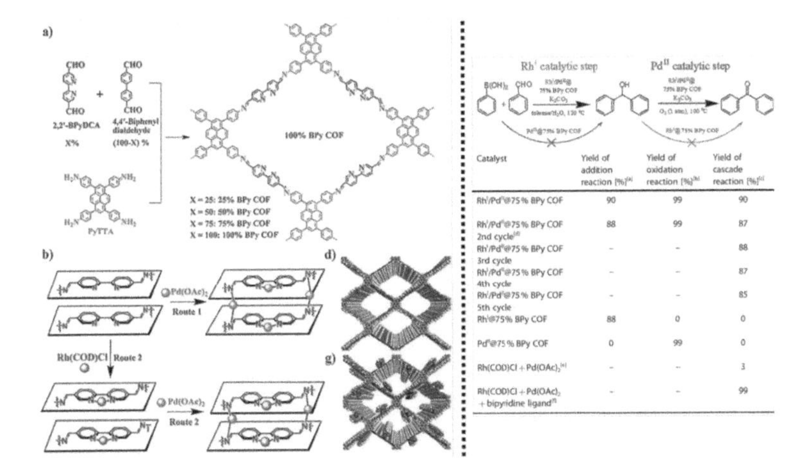

Catalyst	Yield of addition reaction [%][a]	Yield of oxidation reaction [%][b]	Yield of cascade reaction [%][c]
Rh^I/Pd^{II}@75% BPy COF	90	99	90
Rh^I/Pd^{II}@75% BPy COF 2nd cycle[d]	88	99	87
Rh^I/Pd^{II}@75% BPy COF 3rd cycle	-	-	88
Rh^I/Pd^{II}@75% BPy COF 4th cycle	-	-	87
Rh^I/Pd^{II}@75% BPy COF 5th cycle	-	-	85
Rh^I@75% BPy COF	88	0	0
Pd^{II}@75% BPy COF	0	99	0
Rh(COD)Cl + $Pd(OAc)_2$[e]	-	-	3
Rh(COD)Cl + $Pd(OAc)_2$ + bipyridine ligand[f]	-	-	99

Figure 4.19 (a) Use of a three-component condensation system to modulate the nitrogen content of the 2D imine-type COFs, (b) designed strategies for the monometallic (Route 1) and bimetallic docking (Route 2), (c) open channels of the COFs, and (d) open channels of the metal-loaded COFs. Right: One-pot cascade reactions using different homogeneous/heterogeneous catalysts. Reproduced with permission from Ref. [28]. Copyright (2016), John Wiley and Sons.

4.4 Heterogeneous Photo- and Electrocatalysts of COFs

Porous network structures, together with robustness and molecular functionality with the possibility to load desirable species, favor the generation of highly effective photocatalysts and electrocatalysts. Lotsch et al. reported a new COF capable of visible-light-driven hydrogen generation in the presence of Pt as a proton reduction catalyst (PRC) [29]. A hydrazone-linked COF, 1,3,5-tris(4-formylphenyl)triazine (TFPT) COF, was synthesized via the acetic acid–catalyzed reversible condensation of TFPT and 2,5-diethoxy-terephthalohydrazide as the building blocks in dioxane/mesitylene at 120°C under solvothermal conditions (Fig. 4.20a). The COF adopts a layered structure with a honeycomb-type lattice featuring mesopores of 3.8 nm and the highest surface area among all

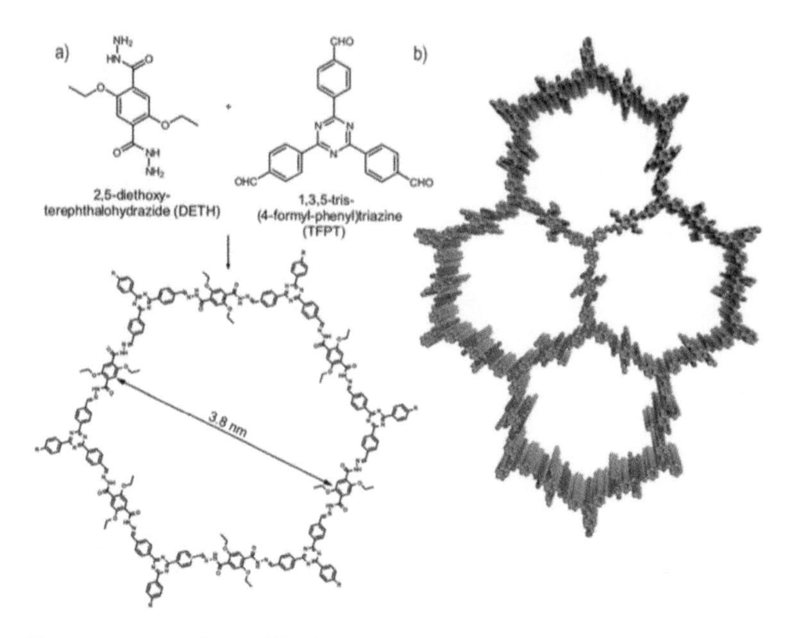

Figure 4.20 Synthesis of hydrazone-linked COF (TFPT COF) with honeycomb mesopores. (a) Scheme showing the condensation of the two monomers to form the TFPT COF and (b) top view with an eclipsed primitive hexagonal lattice (gray: C; blue: N; red: O). Reproduced from Ref. [29] under a Creative Commons Attribution 3.0 Unported Licence (https://creativecommons.org/licenses/by/3.0/).

hydrazone-based COFs (Fig. 4.20a). When illuminated with visible light, the Pt-doped COF continuously produces hydrogen from water without signs of degradation. This result indicates that photoactive COFs are well-defined model systems with which to carry out efficient photocatalytic processes.

CdS NPs have also been deposited on a highly stable 2D COF matrix, and the generated hybrid was used as a photocatalyst for visible-light-driven hydrogen production (Fig. 4.21). In particular, the efficiency of a CdS COF hybrid with different COF contents has been investigated. Upon the introduction of only 1 wt% of COF, a 10-fold increase in the overall photocatalytic activity has been observed. The hybrid with 10 wt% of COF content exhibits a marked H_2 production of approximately 3700 μmol h^{-1} g^{-1}, which is significantly higher than that for bulk CdS NPs (124 μmol h^{-1} g^{-1}) [30]. This study means an effective amalgamation of organic and inorganic materials within a single hybrid, resulting in an improved activity compared to that of each of the constituents.

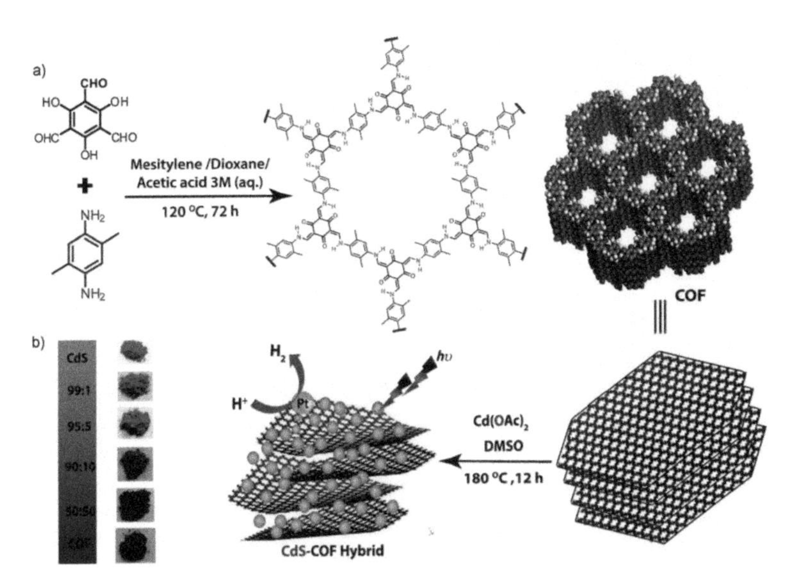

Figure 4.21 Representation of (a) synthesis of the COF TpPa-2. (b) CdS COF hybrid formation by the hydrothermal synthesis of CdS nanoparticles on the COF matrix. Reproduced with permission from Ref. [30]. Copyright (2014), John Wiley and Sons.

A series of water- and photostable 2D azine-linked COFs have been synthesized from hydrazine and triphenylarene aldehyde with a varying number of nitrogen atoms [31]. The electronic and steric variations in the precursors are transferred to the resulting frameworks, leading to a progressively enhanced light-induced hydrogen evaluation with increasing nitrogen content in the frameworks. These results demonstrate that by the rational design of COFs on a molecular level, it is possible to precisely adjust their structural and optoelectronic properties, resulting in enhanced photocatalytic activities. In addition, structure-property-activity relationships in a pyridine unit containing an azine-linked COF for photocatalytic hydrogen evolution have been elucidated [32]. Owing to their inherent porosity and well-ordered nanoscale architectures, COFs are an especially attractive platform to tune photocatalytic hydrogen evaluation, which is extended to a pyridine-based photocatalytic active framework, where nitrogen substitution in the peripheral aryl rings reverses the polarity compared to the above description (Ref. [31]). Lotsch et al. demonstrated how simple changes at the molecular level translate into significant differences in atomic-scale structure, nanoscale morphology, and optoelectronic properties, which greatly affect the photocatalytic hydrogen evolution efficiency. In an effort to understand the complex interplay of such factors, the authors carved out the conformational flexibility of the PTP COF precursor and the vertical radical anion stabilization energy as important descriptors to understand the performance of COF photocatalysts. A benchmark example of COF-based photocatalysts for solar fuel production from CO_2 has been realized. It is demonstrated that the visible-light-harvesting capacity, suitable bandgap, and highly ordered π electron channels contribute to the excellent performance of the COF film photocatalyst [33]. In 2013, Jiang et al. first examined the photocatalytic activity of squaraine-linked cupper porphyrin (CuP) COF by using 1,3-diphenylisobenzofuran (DPBF) as a label for singlet oxygen generation [34]. Further in 2015, Jiang et al. also investigated the photocatalytic activity of an imine-linked CuP COF by using DPBF for singlet oxygen generation. In comparison with other CuP derivatives, the CuP-dihydroxyphenyl (DHPh) COF is exceptionally active as a photocatalyst, exhibiting a 10- to 20-fold enhancement in activity [35].

Recently, a π-conjugated TpMA COF, which was composed of TFP and melamine (MA), with triazine units and cyclic ketone units was artfully designed and synthesized. This COF was found to exhibit an excellent visible-photocatalytic capacity for the decomposition of organic pollutants. The triazine units that were artfully integrated into the COF skeleton served as photoactive centers, and the cyclic ketone units served as electron-withdrawing moieties. Therefore, the conjugated structure served as a photoelectron shift platform [36]. Chemically stable CTFs have been also used as photocatalysts for H_2 evolution in the presence of Pt under visible-light irradiation, and relatively low hydrogen evolution rates of approximately 200 μmol h^{-1} g^{-1} were obtained [37]. Subsequently, a fast and facile route for the optimization of CTFs for photocatalytic hydrogen production was presented by Thomas and coworkers [38]. They found that the optimized CTF catalysts showed an average hydrogen evolution rate of 1072 μmol h^{-1} $g{-}^{1}$ under visible light (>420 nm). Very recently, Lotsch et al. further demonstrated photocatalytic hydrogen evolution using COF photosensitizers with molecular PRCs.

Using an azine-linked COF (N_2) photosensitizer, a chloro(pyridine) cobaloxime cocatalyst, and a triethanolamine donor, a H_2 evolution rate of 782 mmol h^{-1} g^{-1} and a TON of 54.4 was obtained in a water/acetonitrile mixed solvent [39]. In addition to hydrogen evolution, a 2D COF was found to be a highly efficient, metal-free recyclable heterogeneous photocatalyst for oxidative C–H functionalization under visible-light irradiation using O_2 as a green oxygen source [40]. In addition, 2D COFs can act as efficient type II photosensitizers for photodynamic inactivation of bacteria [41].

4.5 Heterogeneous Catalysts of 3D COFs

In 2014, Yan and coworkers designed and prepared two 3D microporous base–functionalized COFs, termed BF-COF-1 and BF-COF-2, via the condensation reaction of a tetrahedral alkyl amine, 1,3,5,7-tetraaminoadamantane (TAA), combined with 1,3,5-triformylbenzene (TFB), or TFP, as shown Fig. 4.22a [42]. Using these two COFs as catalysts, both BF COFs showed remarkable conversion rates (96% for BF-COF-1 and 98% for BF-COF-2) and good recyclability in base-catalyzed Knoevenagel condensation reactions. Importantly, the COFs exhibited highly efficient size selectivity due to the pore size effect of COFs (Fig. 4.22b). The strong interaction between reactants and the COFs may lead to a high efficiency in size selectivity.

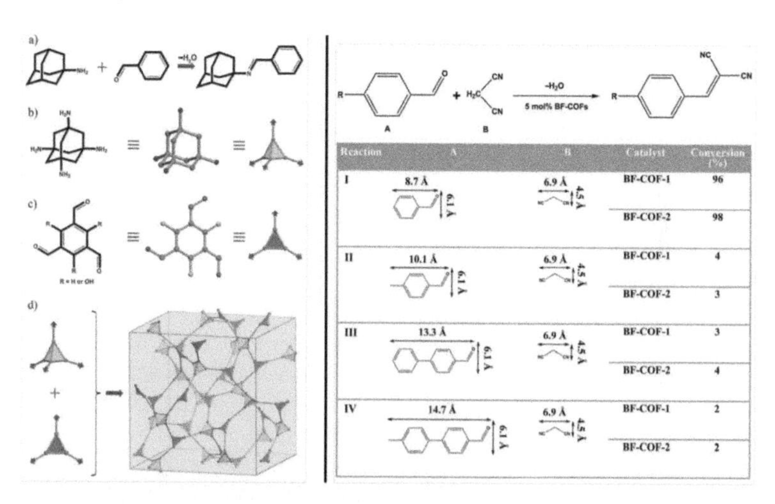

Figure 4.22 Left: Schematic representation of the strategy for preparing 3D microporous base–functionalized COFs. (a) Model reaction of 1-adamantanamine with benzaldehyde to form the molecular *N*-(1-adamantyl) benzaldehyde imine, (b) structure of 1,3,5,7-tetraaminoadamantane (TAA) as a tetrahedral building unit, (c) structure of 1,3,5-triformylbenzene (R = H, TFB) or triformylphloroglucinol (R = OH, TFP) as a triangular building unit, and (d) condensation of tetrahedral and triangular building units to give a 3D network with the symbol ctn (BF-COF-1 and BF-COF-2). Right: Catalytic activity of BF COFs in the size-selective Knoevenagel condensation reaction. Reproduced with permission from Ref. [42]. Copyright (2014), John Wiley and Sons.

Recently, Wang and coworkers described that an interpenetrating dynamic 3D COF LZU-301 could be a Lewis-base catalyst for the Knoevenagel condensation between malonitrile and three aromatic aldehydes, as shown in Fig. 4.23 [43]. The authors discovered that for the small-sized aldehyde, the 3D COF LZU-301 provided an excellent yield of up to 72% in 4 h and 99% in 10 h. However, for the larger 2-napthalenealdehyde and 9-anthracenealdehgyde, the yield notably decreased to 21% and 5% because of a size effect of the pore. The larger fuse-ring aromatic aldehyde could not be accommodated into the pore space of LZU-301 and thus did not have enough interaction with its pyridine Lewis-base catalytic site, leading to inferior catalytic performance. This example manifests its significance in that LZU-301 was one of the few 3D COFs that demonstrate considerable catalytic activities in meaningful organic synthesis.

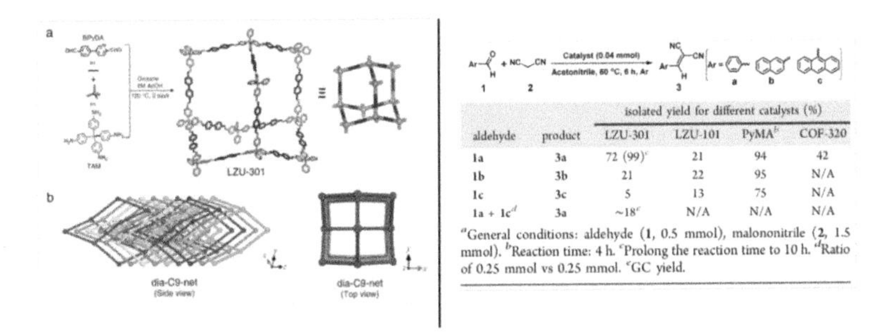

aldehyde	product	isolated yield for different catalysts (%)			
		LZU-301	LZU-101	PyMA[b]	COF-320
1a	3a	72 (99)[c]	21	94	42
1b	3b	21	22	95	N/A
1c	3c	5	13	75	N/A
1a + 1c[d]	3a	~18[e]	N/A	N/A	N/A

[a]General conditions: aldehyde (1, 0.5 mmol), malononitrile (2, 1.5 mmol). [b]Reaction time: 4 h. [c]Prolong the reaction time to 10 h. [d]Ratio of 0.25 mmol vs 0.25 mmol. [e]GC yield.

Figure 4.23 (a) Solvothermal synthesis of a 3D COF, LZU-301, via imine condensation. For clarity, only the single framework of LZU-301 is shown. (b) Side and top views of the porous crystalline structure of LZU-301, which features a ninefold interpenetration of the underlying diamond net. Different colors represent different penetrating frameworks from a side view under reaction condensation. Reprinted with permission from Ref. [43]. Copyright (2017) American Chemical Society.

The 3D COFs DL-COF-1 and DL-COF2 were prepared from the dual linking reaction between TAA and 4-formylphenylboronic acid (FPBA) or 2-fluoro-4-formylphenylboronic acid (FFPBA), forming two kinds of linkages in the COF skeleton, boroxine and imine bonds (Fig. 4.24) [44]. The as-synthesized 3D COFs displayed large specific surface areas and incorporated both acidic boroxine sites and basic imine sites. These two different sites rendered these 3D COFs as versatile bifunctional heterogeneous catalysts. To demonstrate the catalytic activity of DL-COF-1, a one-pot deacetalization-Knoevenagel reaction was applied. DL-COF-1 exhibited excellent yields in both acid-catalyzed deacetalization reaction (yield up to 100%) and base-catalyzed Knoevenagel condensation reaction (yield up to 98%). The COF crystals can be recycled and reused three times with almost no loss of activity and no identical change in PXRD and H_2 uptake characterization. This experiment was very encouraging since it produced the first bifunctional 3D COF with a large surface area to fulfill heterogeneous catalytic applications.

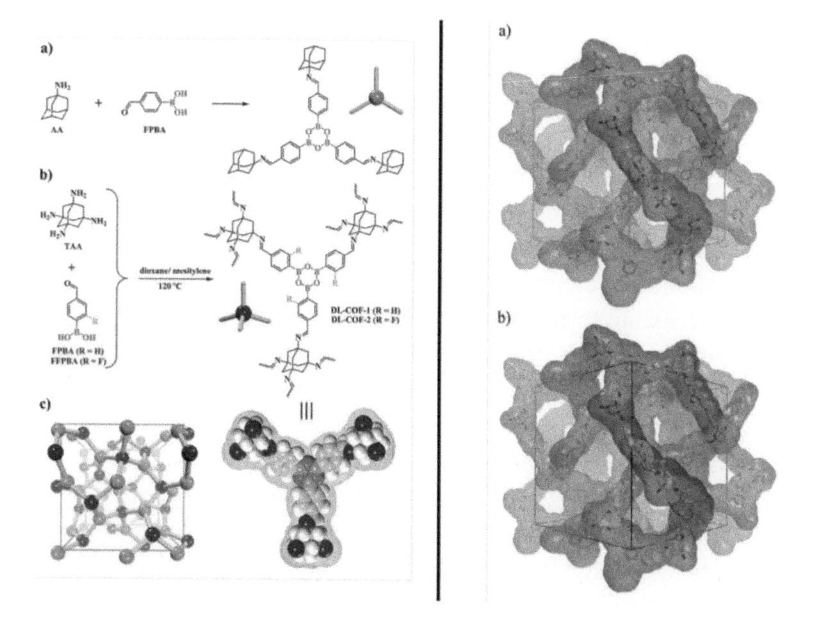

Figure 4.24 Left: Strategy for synthesizing 3D COFs with dual linkages (DL-COF), (a) model reaction of AA with FPBA to form a triangular molecule with dual linkages, (b) condensation of tetrahedral TTA and FFPBA to give 3D COFs with dual linkages, DL-COF-1 or DL-COF-2, and (c) on the basis of triangular and tetrahedral building units, both DL COFs show 3D networks with a ctn topology. Right: Structural representations of 3D DL-COF-1 (a) and DL-COF-2 (b). Reprinted with permission from Ref. [44]. Copyright (2016) American Chemical Society.

4.6 Conclusions and Outlook

Owing to the unique features of COFs and recent interesting exploitation of their properties, COFs undoubtedly demonstrate significant potential and advantages as catalysts. Therefore, COF catalysis has attracted a great deal of attention, and some progress in the field has been made in recent years. However, the investigation of catalytic COF materials is still in its infancy, and COF catalysis still faces many challenges in practical applications, for example, how to further improve the chemical/thermal stability of COFs so that more reactions can be carried out under very harsh conditions (e.g., strong acidity or alkalinity, high temperature, and high overpotential). The preparation of successful crystalline COFs has long been an intractable problem because there is a lack of universal regulation to construct crystalline COFs and significant time and effort must be spent on searching for the appropriate reaction conditions. Therefore, there is a clear need to expand the synthetic possibilities in different ways. Currently, the synthesis of COFs is limited to the laboratory and it has been a challenge to prepare COFs on an industrial scale for prospective practical applications. Development of 3D COFs with versatile and cavities is very interesting to provide active sites on the pore surface and to transport reactants/products for formation of inner reactive vessels. This would enable supramolecular catalysis or enzyme catalysis to take place in an appropriate space under nanoscale confinement, allowing multicomponent reactions and a wide range of substrates from small molecules to large polyaromatics and carbohydrates.

In the future, the goal of COF catalysis is to develop heterogeneous catalysts with high stability, low cost, and high conversion and selectivity. New synthesis methods and process routes for large-scale synthesis should be explored simultaneously, thus laying the foundation for the commercialization and industrialization of COF catalysts.

References

1. Beletskaya, I. P., Cheprakov, A. V. (2000). *Chem. Rev.*, **100**, 3009–3066.
2. Takemoto, T., Iwasa, S., Hamada, H., Shibatomi, K., Kameyama, M., Motoyama, Y., Nishiyama, H. (2007). *Tetrahedron Lett.*, **48**, 3397–3401.
3. Ozawa, F., Kubo, A., Hayashi, T. (1993). In *Selectivity in Catalysis*; ACS Symposium Series 517; ACS: Washington, DC, USA, Vol. 517, 75–85.
4. Kamei, T., Sato, A. H., Iwasawa, T. (2011). *Tetrahedron Lett.*, **52**, 2638–2641.
5. Liang, J., Liang, Z., Zou, R., Zhao, Y. (2017). *Adv. Mater.*, **29**, 1701139.
6. Ding, S.-Y., Gao, J., Wang, Q., Zhang, Y., Song, W.-G., Su, C.-Y., Wang, W. (2011). *J. Am. Chem. Soc.*, **133**, 19816–19822.
7. Hou, Y., Zhang, X., Sun, J., Lin, S., Qi, D., Hong, R., Li, D., Xiao, X., Jiang, D. (2015). *Microporous Mesporous Mater.*, **214**, 108–114.
8. Llabres i Xamena, F. X., Abad, A., Corma, A., Garcia, H. (2007). *J. Catal.*, **250**, 294–298.

9. Mullangi, D., Nandi, S., Shalini, S., Sreedhala, S., Vinod, C. P., Vaidhyanathan, R. (2015). *Sci. Rep.*, **5**, 10876.
10. Kaleeswaran, D., Antony, R., Sharma, A., Malani, A., Murugavel, R. (2017). *Chempluschem*, **82**, 1253–1265.
11. Lu, S., Hu, Y., Wan, S., McCaffrey, R., Jin, Y., Gu, H., Zhang, W. (2017). *J. Am. Chem. Soc.*, **139**, 17082–17088.
12. Beletskaya, I. P., Cheprakov, A. V. (2000). *Chem. Rev.*, **100**, 3009–3066.
13. Pachfule, P., Panda, M. K., Kandambeth, S., Shivaprasad, S. M., Diaz Daiz, D., Banerjee, R. (2014). *J. Mater. Chem. A*, **2**, 7944–7952.
14. Zhang, J., Peng, Y., Leng, W., Gao, Y., Xu, F., Chai, J. (2016). *Chin. J. Cata.*, **37**, 468–475.
15. Bhadra, M., Sasmal, H. S., Basu, A., Midya, S. P., Kandambeth, S., Pachfule, P., Balaraman, E., Banerjee, R. (2017). *ACS Appl. Mater. Interface*, **9**, 13785–13792.
16. Lin, S., Hou, Y., Deng, X., Wang, H., Sun, S., Zhang, X. (2015). *RSC Adv.*, **5**, 41017–41024.
17. Nagai, A., Guo, Z., Feng, X., Jin, S., Chen, X., Ding, X., Jiang, D. (2011). *Nat. Commun.*, **2**, 536, doi: 10.1038/ncomms1542.
18. Xu, H., Chen, X., Gao, J., Lin, J., Addicoat, M., Irle, S., Jiang, D. (2014). *Chem. Commun.*, **50**, 1292–1294.
19. Xu, H., Cao, J., Jiang, D. (2015). *Nat. Chem.*, **7**, 905–912.
20. Ma, H. C., Kan, J. L., Chen, G. J., Chen. C. X., Dong, Y. B. (2017). *Chem. Mater.*, **29**, 6518–6524.
21. Zhang, J., Han, X., Wu, Z. W., Liu, Y., Chu, Y. (2017). *J. Am. Chem. Soc.*, **139**, 8277–8285.
22. Han, X., Xia, Q., Huang, J., Liu, Y., Tan, C., Cui, Y. (2017). *J. Am. Chem. Soc.*, **139**, 8693–8697.
23. Peng, Y. W., Hu, Z. G., Gao, Y. J., Yuan, D. Q., Kang, Z. X., Qian, Y. H., Yan, N., Zhao, D. (2015). *ChemSusChem*, **8**, 3208–3212.
24. Wang, X. R., Han, X., Zhang, J., Wu, X. W., Liu, Y., Cui, Y. (2016). *J. Am. Chem. Soc.*, **138**, 12332–12335.
25. Xu, H. S., Ding, S. Y., An, W. K., Wu, H., Wang, W. (2016). *J. Am. Chem. Soc.*, **138**, 11489–111492.
26. Shinde, D. B., Kandambeth, S., Pachfule, P., Kumar, R. R. (2015). *Chem. Commun.*, **51**, 310–313.
27. Leng, W., Ge, R., Dong, B., Wang, C., Gao, Y. (2016). *RSC Adv.*, **6**, 37403–37406.

28. Leng, W., Peng, Y., Zhang, J., Lu, H., Feng, X., Ge, R., Dong, B., Wang, B., Gao, Y. (2016). *Chem. Eur. J.*, **22**, 9087–9091.
29. Stegbauer, L., Schwinghammer, K., Lotsch, B. V. (2014). *Chem. Sci.*, **5**, 2789–2793.
30. Thote, J., Aiyappa, H. B., Deshpande, A., Díaz, D. D., Kurungot, S., Banerjee, R. (2014). *Chem. Eur. J.*, **20**, 15961–15965.
31. Vyas, V. S., Haase, F., Stegauer, L., Savasci, G., Podjaski, F., Ochenfeld, C., Lotsch, B. V. (2015). *Nat, Commun.*, **6**, 8508.
32. Haase, F., Banerjee, T., Savasci, G., Christian, O., Lotsch, B. V. (2017). *Faraday Discuss.*, **201**, 247–264.
33. Yadav, R. K., Kumar, A., Park, N. J., Kong, K. J., Baeg, J. O. (2016). *J. Mater. Chem. A*, **4**, 9413–9418.
34. Nagai, A., Chen, X., Feng, X., Ding, X., Guo, Z., Jiang, D. (2013). *Angew. Chem. Int. Ed.*, **52**, 3770–3774.
35. Chen, X., Addicoat, M., Jing, E. Q., Zhai, L. P., Xu, H., Huang, N., Guo, Z. Q., Liu, L. L., Irle, S., Jiang, D. L. (2015). *J. Am. Chem. Soc.*, **137**, 3241–9639.
36. He, S., Rong, Q., Niu, H., Cai, Y. (2017). *Chem. Commun.*, **53**, 9636–9639.
37. Bi, J., Fang, W., Li, L., Wang, J., Liang, S., He, Y., Liu, M., Wu, L. (2015). *Macromol. Rapid Commun.*, **36**, 1799–1805.
38. Kuecken, S., Acharjya, A., Schwarze, M., Schomäcker, R., Thomas, A. (2017). *Chem. Commun.*, **53**, 5854–5857.
39. Banerjee, T., Hasse, F., Savasci, G., Gottschling, K., Ochsenfeld, C., Lotsch, B. V. (2017). *J. Am. Chem. Soc.*, **139**, 16228–16234.
40. Zhi, Y., Li, Z., Feng, X., Xia, H., Zhang, Y., Shi, Z., Mu, Y., Liu, X. (2017). *J. Mater. Chem. A*, **5**, 22933–22938.
41. Liu, T., Hu, X., Wang, Y., Meng, L., Zhou, Y., Zhang, J., Chen, M., Zhang, X. (2017). *J. Photochem. Photobiol. B*, **175**, 156–162.
42. Fang, Q. R., Gu, S., Zheng, J., Zhuang, Z. B., Qiu, D. L., Yan, Y. S. (2014). *Angew. Chem. Int. Ed.*, **53**, 2878–2882.
43. Ma, Y. X., Li, Z. J., Wei, L., Ding, S. Y., Zhang,Y. B., Wang, W. (2017). *J. Am. Chem. Soc.*, **139**, 4995–9091.
44. Li, H., Pan, Q., Ma, Y., Guan, X., Xue, M., Fang, Q., Yan,Y., Valtchev, V., Qiu, S. (2016). *J. Am. Chem. Soc.*, **138**, 14783–14788.

Chapter 5

Energy Storage Applications of 2D COFs

Synthetic polymers with branched macromolecules and outstanding functional group tolerance show diverse and useful properties that influence most aspects of modern life. Extending polymerization strategies to two dimensions allows precise integration of building blocks into extended structures with periodic skeletons and ordered nanopores. The construction principle of these frameworks is the direct topological evaluation in a predictable manner with controlled geometry, dimensions, and structural periodicity. This unique designable feature of 2D covalent organic frameworks (COFs) with versatile properties makes them an emerging material platform with great relevance in areas such as gas storage and separation (described in Chapter 3), catalysis (described in Chapter 4), and optoelectronics. This chapter focuses on the recent progress in 2D COFs as optoelectronic materials, with an emphasis on their semiconducting, energy conversion, and energy storage properties.

5.1 2D COFs for Optoelectronics and Energy Storage

2D covalent polymers, such as covalent organic frameworks (COFs), in which building blocks are precisely integrated into extended structures with periodic skeletons and ordered pores, have some

Covalent Organic Frameworks
Atsushi Nagai

ISBN 978-981-4800-87-7 (Hardcover), 978-1-003-00469-1 (eBook)
www.jennystanford.com

distinctive advantages [1–3]. The unique topological diagram directs the growth of the frameworks in a predictable manner, and the geometry and dimensions of the building blocks govern the structures of the resulting COFs. In 2D COFs, building units are stacked via π-π interactions to form layered structures with a well-defined alignment of π building units to their atomic layers and segregated arrays of π columns. In such arrays, the intralayered covalent bonds lock the frameworks whereas interlayer noncovalent interaction controls the stability. Thus, various COFs can be achieved by controlling pore design from a material-design perspective, by controlling the skeleton design, or by using complementary designs of both pores and skeleton. Such varying designable features of 2D COFs, with confined spaces in controllable 1D nanochannels, offer the possibility to trigger interactions with excitons, electrons, holes, spins, ions, and molecules. By this means, 2D COFs exhibit unique properties and functions with outstanding applicability in semiconductors [4], gas absorption [5, 6], proton conduction [7, 8], luminescence [9, 10], and energy conversion and storage.

5.2 Semiconducting and Photoconducting 2D COFs

Many conducting materials of organic compounds have been designed by fine-tuning their energy gaps and highest occupied molecular orbital/lowest occupied molecular orbital (HOMO/LUMO) levels and can be efficiently produced on a large scale by the versatile tools of organic chemistry [11]. The electronic devices fabricated with these materials often exhibit lower stability and efficiency than expected due to inefficient stacking or disordered donor-acceptor interactions. Thus, it would be desirable to have synthetic access to conducting materials with total control over their nanoscale structures and orientations for the fabrication of nanoscale devices and to expand our understanding of the fundamental physics of 2D semiconducting materials. In this respect, 2D COFs, in which modularly designed organic building blocks are periodically linked with regular pores, offer the possibility of developing fully conjugated 2D crystalline networks. Such periodically stacked alignment develops columns in a direction that favors the transport of charge carriers and photoexcited states (excitons) produced from excited building blocks through preorganized and built-in pathway. This unidirectional charge transport improves the carrier mobility of 2D COFs and provides a new platform for semiconducting and photoconducting materials design. A selection of semiconducting COFs is summarized in Table 5.1.

Table 5.1 Summarized semiconducting property of some COFs

COFs	Semiconducting property	COFs	Semiconducting Property
TP COF	p-type	D_{TP}-A_{NDI} COFs	Ambipolar type
PPy COF	p-type	D_{TP}-A_{PyDI} COFs	Ambipolar type
MPc COFs (M = Ni, Co, Cu, Zn)	p-type	Tp-P COFs	Ambipolar type
COF-366	p-type	ZnP COFs	Ambipolar type
COF-66	p-type	D_{MPc}-A_{PyDI} COFs (M = Cu, Ni)	Ambipolar type
H_2P COFs	p-type	D_{MPc}-A_{NDI} COFs (M = Cu, Ni, Zn)	Ambipolar type
T COFs	p-type	D_{MPc}-A_{PDI} COFs (M = Cu, Zn)	Ambipolar type
TTF-Ph COFs	p-type	C_2N-h_2D	Ambipolar type
TTF-Py COFs	p-type	$[C_{60}]_y$-ZnPC COFs	Ambipolar type
NiPc-BTTA COFs	n-type	$[C_{60}]$-TT COFs	Ambipolar type
CuP COFs	n-type	CS-COF-C_{60}	Ambipolar type
D-A COFs	Ambipolar type		

TP, triphenylene; PPy, pyrene-2,7-diboronic acid; Pc, phthalocyanine; P, porphyrin; T, thiophene; TTF, tetrathiafulvalene; Ph, phenyl; Py, pyrene; BTDA, benzothiazole; D, donor; A, acceptor; PyDI, pyromellitic diimide; NDI, naphthalene diimide; PDI, perylene diimide

5.3 P-Type Semiconducting 2D COFs

In general, COFs with extended p-conjugated systems, designed by using more electron-rich building blocks, display p-type semiconducting behavior. Jiang and coworkers pioneered in this field and first demonstrated the luminesce and semiconducting behavior of COFs consisting of alternating pyrene and triphenylene functionalities (named TP COFs) as shown in Fig. 5.1 [12]. The co-condensation of 2,3,6,7,10,11-hexahydroxytriphenylene (HHTP) and pyrene-2,7-diboronic acid (PDBA) adopts a belt-shaped hexagonal crystalline framework with a pore diameter of 3.14 nm and a specific surface area and a pore volume of 868 m^2/g and 0.7907 cm^3/g, respectively. Because of matching spectral profiles of the building blocks, a TP COF harvests a wide wavelength range of photons, allowing energy transfer and migration between them and blue luminesce. Upon excitation of triphenylene units of a TP COF at 340 nm, two bands were visible, with a weak emission maximum at 402 nm and a strong emission maximum at 474 nm. In contrast, the individual monomers on excitation show emission maxima at 402 nm (triphenyl unit) and 475 nm (pyrene unit). The difference in fluorescence intensities at 474 and 402 nm ($I_{474\ nm}/I_{402\ nm} = 16$) is viewed as evidence of electronic coupling of the building blocks, which favors an energy transfer between them in the extended TP-COF structure. In addition, fluorescence anisotropy measurements illustrate that this excitation energy migrates randomly over the crystalline belt within the lifetime of the excited state as well. To measure the electrical conductivity, a disappeared COF solution in acetone was drop-casted onto a Pt electrode with a width of 10 mm. In air at 25°C the *I-V* profile of the PT COF was almost linear, irrespective of the voltage bias, while the gap itself was silent, showing the semiconducting nature of this material. The stability of this material was shown from continual on-off switching for many cycles without significant weakening of the electric current. Upon doping the framework with iodine, the electric current increased from 4.3 to 20 nA at a bias voltage of 2 V, indicating the p-type semiconducting character of the TP COF. They also reported the continuous self-condensation of PDBA to afford a PPy COF with a pore diameter of 1.73 nm and a surface area of 923 m^2/g (Fig. 5.1) [13]. The PPy COF displayed a photoconducting behavior with an electrical conductivity of the same order as that of a TP COF. The photoconductivity of the PPy COF was investigated using a sandwich electrode, fabricated by casting a thin film (PPy COF/PMMA = 50/50, wt%) with a thickness of 100 mm on an Al electrode and covering the film with a 30 nm thick layer of Au by vapor deposition. When the PPy COF is irradiated with visible light (>400 nm) from the Au side with a xenon lamp, a remarkable photocurrent is generated that can be switched repetitively many times without any deterioration at an on/off ratio of over 8×10^4.

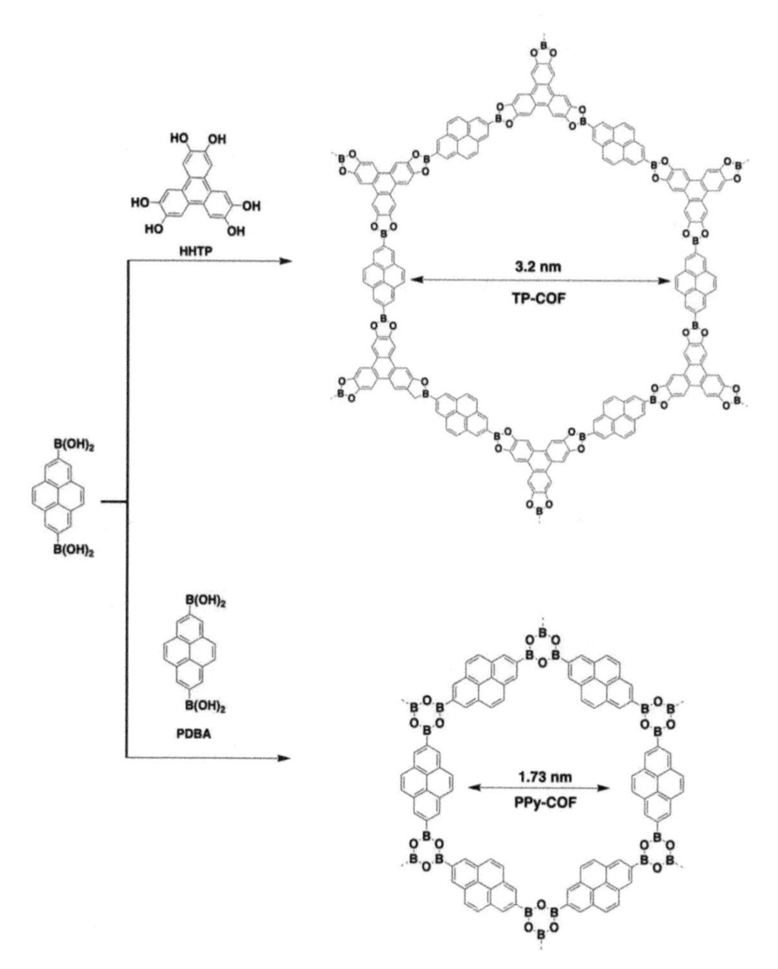

Figure 5.1 Schematic representation of the chemical structures of TP COF and PPy COF [12, 13].

Porphyrin and phthalocyanine are structurally related macrocyclic compounds, ubiquitous in nature and serving as important components of molecular materials with electronic, magnetic, and physicochemical properties. The incorporation of these into 2D COFs, where π electron systems align in a precise order, opens the possibility of developing promising photon drive devices. The amalgamation of large metallophthalocyanine (MPc) π systems into 2D COFs based on a boronate esterification reaction was first reported by Jiang et al. [14]. Under solvothermal conditions the condensation of (2,3,9,10,16,17,23,24-octahydroxyphthalocyaninato)nickel(II) and $[(OH)_8PcNi]$ with 1,4-benzenediboronic acid (BDBA) produced the desired 2D NiPc COF with a 90% yield (Fig. 5.2). The resultant COF exists as a layered structure of planar sheets, with a large surface area of 624 m^2/g and uniform microporous channels with a diameter of 1.9 nm. This well-ordered eclipsed stacking

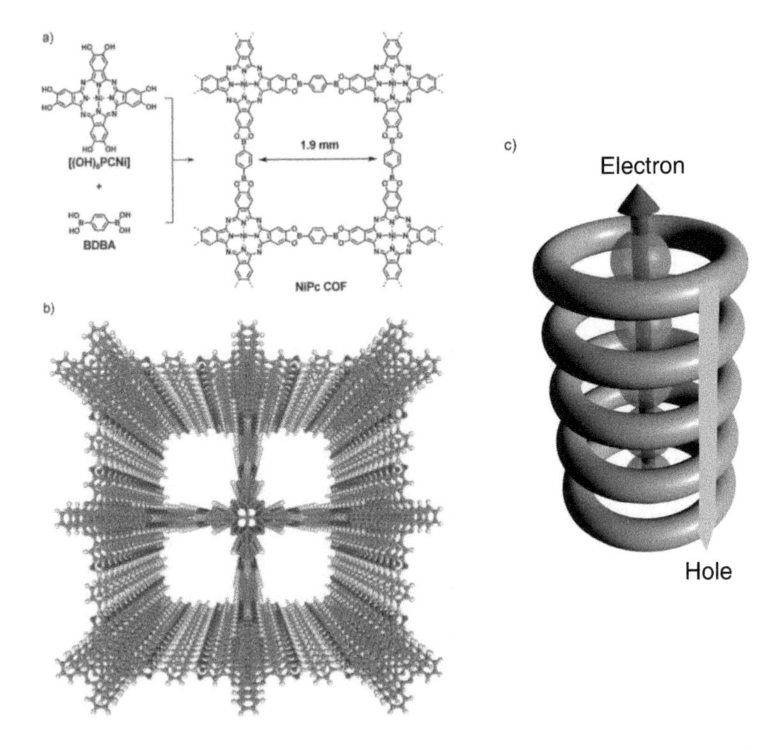

Figure 5.2 (a) Schematic representation of metallophthalocyanine 2D-covalent organic frameworks (MPc COFs), (b) eclipsed stack of phthalocyanine 2D sheets with microporous channels, (c) schematic representation of stacked phthalocyanine π columns in MPc COFs for hole-carrier transport. Reproduced with permission from Ref. [14]. Copyright (2011), John Wiley and Sons.

conformation enhanced its light-harvesting capacity in the deep-red visible and near-infrared (NIR) regions. Upon irradiation with a xenon lamp (>400 nm), a significant increase in current from 20 nA to 3 μA at a 1 V bias was observed, which was reproducible over many cycles without deterioration.

In addition, the wavelength-dependent on-off switching of the photocurrent on irradiation with light passing through band path filters (±5 nm) at the same bias revealed its extreme sensitivity toward deep-red and NIR photons. When sandwiched between Au and Al electrodes, the COF shows linear *I-V* profiles regardless of the bias voltage, indicating its semiconducting nature. Laser flash-photolysis time-resolved microwave conductivity (FP-TRMC) measurements were used to evaluate the intrinsic charge-carrier mobility with the value of $\phi\Sigma\mu$ ($\Sigma\mu$ = the sum of carrier mobility, ϕ = the charge-carrier generation efficiency). The charge-carrier generation efficiency ϕ was determined by the integration of the time-of-flight (TOF) transients at different bias voltages. The charge-carrier efficiency ϕ of 3.0×10^{-5} was estimated by the extrapolation of the bias at 0 V. Persistent results on varying the atmosphere of the measurement signposts indicate that carrier species predominantly originate from holes with a mobility as high as 1.3 $cm^2\ V^{-1}\ s^{-1}$. In a publication that followed, they stretched their search and investigated the role of central ions in controlling the photoelectric function of this COF framework [15]. They expanded the usefulness of phthalocyanine COFs (MPs COFs) with various metal ions (M = Co, Cu, Zn), as shown in Fig. 5.2. In terms of structural interpretation, all the MPc COFs are unique, with highly ordered crystalline skeletons with periodic mesoporous columns. The inherent charge-carrier mobility measured by the FP-TRMC method under identical conditions revealed that a CoPc COF has the highest $\phi\Sigma\mu$ value (of $2.6 \times 10^{-4}\ cm^2\ V^{-1}\ s^{-1}$) followed by a ZnPc COF ($2.2 \times 10^{-4}\ cm^2\ V^{-1}\ s^{-1}$) and a CuPc COF ($1.4 \times 10^{-4}\ cm^2\ V^{-1}\ s^{-1}$). This trend matched well with the reported electron densities of the phthalocyanine macrocycles, in the order CuPc COF < ZnPc COF < CoPc COF [16, 17]. The photocurrent generation upon light irradiation was examined on spin-coating the MPc COFs onto a sandwich-type gap between Al and Au electrodes with an applied bias voltage of 2 V. Although the random monomeric samples are not photoconductive, the MPc COFs show high photoconductivity with a generated photocurrent of approximately 110 nA for a CuPc

COF, followed by 0.6 nA and 0.14 nA for a ZnPc COF and a CoPc COF, respectively. This eccentricity between the order of produced photocurrents and $\phi\Sigma\mu$ values for different MPc COFs highlights the importance of morphologies and boundaries of the objects as well as the role of metal ions to control the performance of photoelectric devices. More recently, two triangular COFs, hexaphenylbenzene (HPB) COFs and hexabenzocoronene (HBC) COFs, were developed by using two different C6-symmetric vertices HPB and HBC based on the triangular topology [18]. This unique topology favors the formation of supermicropores with pore sizes as low as 12 Å and densely packed π columns of density 0.25 nm^{-2}, which exceeds those of the COFs and supramolecular π arrays reported to date. These support intra- and interlayer π cloud delocalization on the framework and display prominent photoconductivity, with carrier mobilities as high as 0.7 cm^2 V^{-1} s^{-1}, which is among the highest reported for 2D COFs and photographitic ensembles.

Yaghi and coworkers reported two unique COFs under solvothermal condensation of porphyrin derivative, either through boronate ester formation with tetrahydroxy anthracene (COF-66) or through imine bond formation with tetraphthaladehyde (COF-366) (Fig. 5.3) [19]. The electrical conductivity of both COFs was examined across a gap of 2 mm between two Au electrodes. In air at room temperature, both COFs display linear *I-V* curve profiles and electrical conductivity with an electric current of 0.75 nA at a 0.2 V bias voltage. The 1.5 mm thin film fabricated from COF/poly(methyl methacrylate), 60/40 in wt%, between Al and indium tin oxide (ITO) electrodes, showed hole conduction on TOF transient current integration measurement. This 1D hole mobility of 8.1 cm^2 V^{-1} s^{-1} and 3 cm^2 V^{-1} s^{-1}, respectively, from imine and boronate ester COFs indicates that both are p-type semiconductors. This is also attributed to the fact that the imine bond in COF-366 improves the conjugation of the framework, responsible for the highest hole mobility in semiconducting COFs up to now. Jiang et al. reported a disk-shaped tetragonal H_2P COF, with a porphyrin unit embedded in the mesoporous framework (Fig. 5.4) [20]. The porphyrin unit is aligned in an eclipsed stacking mode with high crystallinity and a large surface area (1901 m^2/g) and offers macrocycle-on-macrocycle columnar porphyrin paths that favor condition pathways for high-rate charge-carrier mobility. The charge-carrier mobility

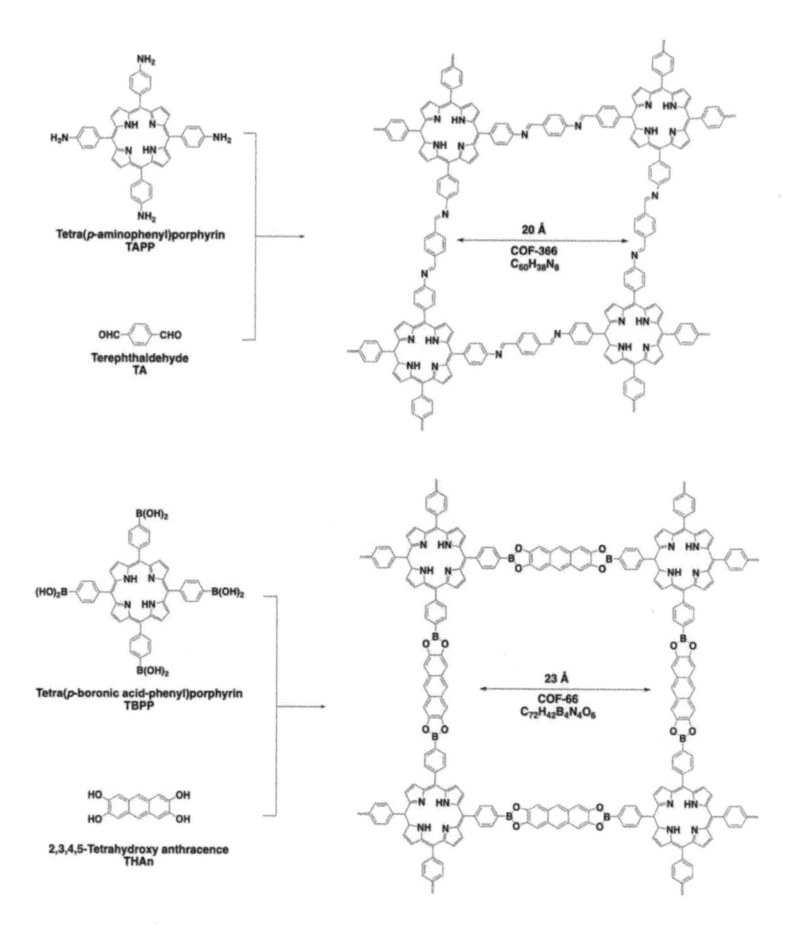

Figure 5.3 Schematic representation of the chemical structures of COF 266 and COF 66.

was evaluated with FP-TRMC methods after laser pulse irradiation under a rapidly oscillating electric field. In both conditions (Ar and SF6 atmosphere) the hole transport mobility displays the same value ($\phi\mu_h = 1.8 \times 10^{-4}$ cm^2 V^{-1} s^{-1}) for the H_2P COF. The maximum quantum yield (ϕ) of photocarrier generation was 5×10^{-5}, measured with the TOF-transient current integration method. This implies a minimum μ_h value of 3.5 cm^2 V^{-1} s^{-1}, which is twofold higher than that of the previously reported phthalocyanine COF (1.6 cm^2 V^{-1} s^{-1}). The photocurrent generation of approximately 0.01 nA was measured upon irradiation with visible light (>400 nm) using a Xe lamp with an on-off ratio of 4. The unbalanced carrier (hole and electron) transport was responsible for low photocurrent generation as well as low on-off ratio.

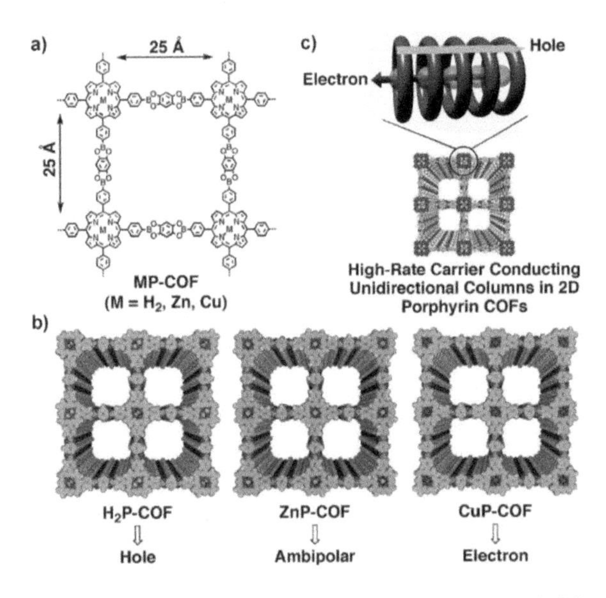

Figure 5.4 (a) Chemical structures of MP COFs (M = H_2, Zn, Cu), (b) schematic graphs of metal-on-metal and macrocycle-on-macrocycle channels for electron transport and hole transport, respectively, in stacked porphyrin columns of 2D MP COFs, (c) schematic graphs of MP COFs with achiral AA stacking of 2D sheets (C: light blue; N: deep blue; H: white; O: red; B: pink; Zn: green; Cu: violet). Reproduced with permission from Ref. [20]. Copyright (2012), John Wiley and Sons.

In the recent past, thiophene- and tetrathiafulvalene (TTF)-based derivatives have been frequently utilized as strong electron-donor molecules for the development of electrically conducting materials [21]. This has led different research groups to use these derivatives as basic building blocks for the synthesis of semiconducting COFs. Dincã et al. isolated the first charge transfer (CT) complex postsynthetically inside a COF embedded with thiophene, bithiophene, and thienothiophene monomers to achieve electric conductivity [22]. The condensation of HHTP with diboronic acid of different thiophene derivatives in dioxane and mesitylene was carried out to afford different thiophene-based COFs (T COFs) as off-white powders with a good yield (Fig. 5.4). All these T COFs preferred to adopt an eclipsed stacking conformation with pore diameters of 2.04, 3.14, and 2.94 nm and surface areas of 927, 544, and 904 m^2/g for T-COF-1, T-COF-3, and T-COF-4, respectively. They performed the doping experiment with these COF frameworks with a suitable acceptor to provide a sufficient concentration of CT to induce electric conductivity. A series of acceptor molecules were examined in search for proper redox partners that promote full electron transfer without disrupting the crystalline structure. Among these 2,3-dichloro-5,6-dicyano-1,4-benzoquinone, chloranil, and I_2 formed the CT complex with these COFs but led to the amorphization of the frameworks. On the other hand, a 0.1 mM solution of tetracyano quinodimethane (TCNQ) in CH_2Cl_2 gave no reaction with T-COF-1 or T-COF-3. It immediately formed a crystalline back precipitate with an adsorption maximum centered at 850 nm with retained crystallinity in contract with solid T-COF-4. Thus, thiophene derivatives with lower oxidation potentials may offer better p-type hosts to form the CT complex in such COF frameworks for the implementation of their materials in electronic applications. Even more recently, Dincã et al. extended their research and reported heavier chalcogen-based COFs in the form of fused benzodiselenophenes and benzoditellurophenes, to compare the electrical properties relative to the thiophene analogue (Fig. 5.5) [23]. Two-point probe electrical conductivity measurements of these COFs revealed values of $3.7 \pm 0.4 \times 10^{-10}$ S/cm, $8.4 \pm 3.8 \times 10^{-9}$ S/cm, and $1.3 \pm 0.1 \times 10^{-17}$ S/cm for S, Se, and Te derivatives, respectively. The enhanced orbital overlap afforded by the 3p and 4p orbitals of selenium and tellurium atoms or the enhanced spin-orbit coupling effects responsible for this trend of electrical conductivity increase in the order of S > Se > Te, without significantly changing the structure or the unit cell size of the resulting extended networks.

More than a few research groups used the important electroactive TTF building block as a basic building unit to fabricate semiconducting COFs. The $C_2 + C_2$ topological diagram to embed TTF units in COF architectures with varying linker units was used by Jiang et al. TTF-based COFs TTF-Ph COF and TTF-Py COF were prepared using phenyldiamine and tetra(*p*-aminophenyl)-pyrene as linkers under solvothermal conditions (Fig. 5.6a) [24]. Both COFs exist as layered lattices with periodic TTF columns and tetragonal open nanochannels. Because of differences in the shape

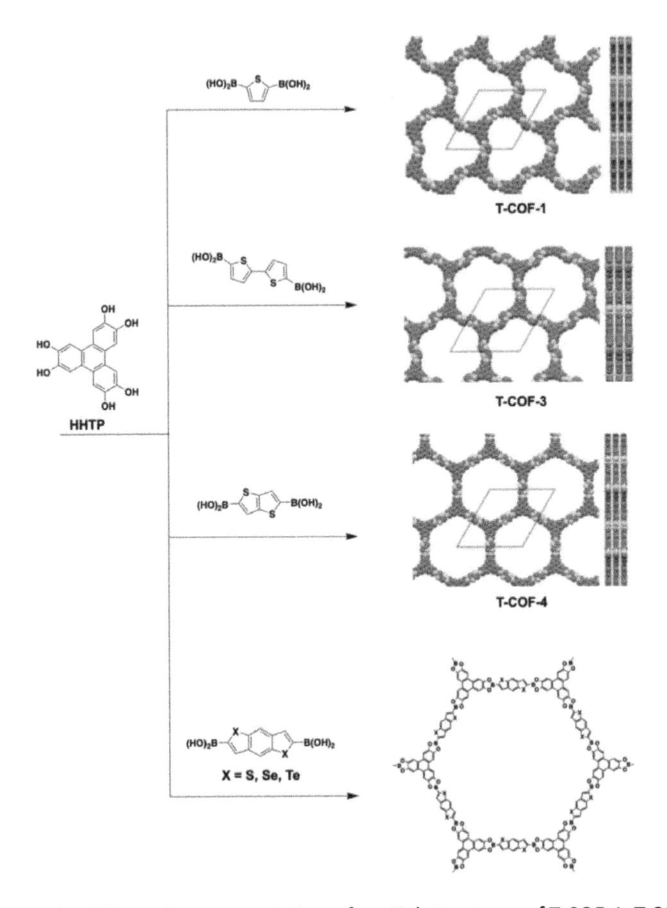

Figure 5.5 Schematic representation of partial structures of T-COF-1, T-COF-3, T-COF-4, and heavier chalcogen-based COFs. Reproduced with from Ref. [22] with permission from PNAS.

of the linkers, the phenyl units adopt a planar conformation with an interlayer distance of 3.71 Å on the TTF-Ph COF, whereas the paddle-shaped tetraphenylpyrene linker distorts the layer from a planar conformation and expands the interlayer distance up to 3.87 Å. Thus, the shape of the linker plays an important role in tuning the periodicity and conformation of the 2D layer with the interlayer distance, which are key to electric functions, such as carrier transport and conductivity. The inherent charge-carrier mobility measured by using the FP-TRMC method after 355 nm laser pulse irradiation revealed values of 0.2 and 0.08 $cm^2\ V^{-1}\ s^{-1}$ for the TTF-

Ph COF and the TTF-Py COF, respectively. Furthermore, conductivity measurements of the COFs were also performed on these 0.4 nm wide and 0.05 nm thick COF films after doping them with iodine vapor. The conductivities of the TTF-Ph COF and the TTF-Py COF were estimated to be 10^{-5} and 10^{-6} Ω^{-1} cm^{-1}, respectively. This higher conductivity value on doping suggests that layered TTF columns in the COF frameworks allow the formation of a CT complex between TTF and the radical cation and the iodide radical anion without disturbing the lattice structure. In addition, the difference in the degrees of enhancement for the TTF-Ph COF and the TTF-Py COF is likely related to the linker nature, which governs the layer conformation and the interlayer distance.

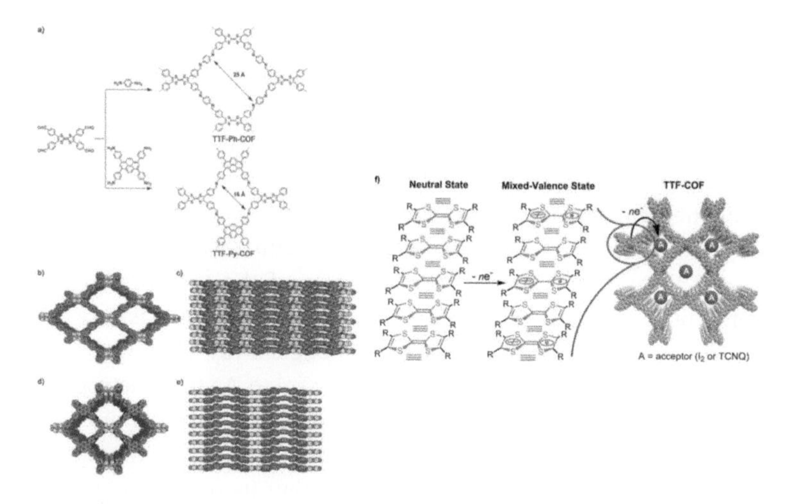

Figure 5.6 (a) Schematic representation of the synthesis of mesoporous TTF-Ph COF and microporous TTF-Py COF by a C_2 + C_2 topological diagram. Slipped AA stacking structures of TTF-Ph COF in (b) top and (c) side views and eclipsed stacking structures of TTF-Py COF in (d) top and (e) side views (yellow: S; blue: N; gray: C; H was omitted for clarity). (f) Illustration of the mixed-valence state in TTF COF. The "h" indicates inter-TTF-layer interactions. Reproduced with permission from Refs. [24]. Copyright (2014), John Wiley and Sons. Republished with permission of Royal Society of Chemistry, from Ref. [26], copyright (2014); permission conveyed through Copyright Clearance Center, Inc.

The synthesis of the TTF-Ph COF using cyclic voltammetry (CV) measurements was investigated in detail for its electrochemical behavior [25]. The working electrode was modified due to low solubility of this COF in common organic solvents. The TTF-Ph COF and carbon black in a ratio of 3:2 wt% were ground in an agate mortar and pestle for 10 min. and sonicated in CH_2Cl_2. Then, the sonicated solution was drop-casted onto the tip of the Pt working electrode. The electron-donating nature of the TTF-Ph COF was established from the CV profile with two reversible redox processes at 0.69 V and 1.07 V. A shape peak at 2.0094, corresponding to the TTF radical cation in the electron paramagnetic resonance (EPR) spectrum, validated the formation of a CT complex upon I_2 adsorption on the COF pore. The electrical conductivity of its doped state contracts in a constant-voltage two-probe configuration. The conductivity value at 25°C increased from 2.1×10^{-7} S/cm to 1.8×10^{-6} S/cm on increasing the doping time from one to two days, indicating that the conducting property of the TTF-Ph COF could be tuned by iodine doping. Zhang et al. reported the oriented thin film of a TTF-Ph COF with an aim to enhance and tune its electrical conductivity in the presence of a molecular dopant as a CT partner [26]. The thin films of the TTF-Ph COF were grown in situ from the liquid phase with a nominal thickness of around 150 nm on Si/SiO_2 substrates and transparent indium thin oxide (ITO)-coated glass. The ordered polycrystalline nature of the thin film with a preferential orientation of the columns normal to the substrate was confirmed from grazing incidence wide-angle X-ray scattering. Thin films on a Si/SiO_2 substrate with prefabricated Au/Cr electrodes 3 mm long, 50 nm thick, and 125 nm apart determine the conductivity of 1.2×10^{-4} S/cm. On exposing this thin film to I_2 vapor in a closed chamber, a significant time-dependent increase in electrical conductivity with a maximum value of 0.28 S/cm occurred after 24 h. Similar conductivity enhancement, albeit of a lower magnitude, was also observed when the film was doped with TCNQ. This highly conductive nature is due to the formation of a CT complex with I_2 or TCNQ and correlates well with diagnostic signatures in ultraviolet-visible-NIR and EPR spectra. Thin films on an ITO transparent glass substrate were used to record the optical adsorption spectra in transmission mode. The low-intensity NIR peak at 1200 nm of the COF film progressively redshifts up to 1400 nm and 2000 nm over the course of TCNQ and I_2 exposure. These NIR adsorption bands signify a mixed-valance TTF species formation between TTF and TTF radical cation, where effective radical delocalization within mixed-valance TTF stacks surges upward upon the formation of a CT complex (Fig. 5.6f). A small amount of paramagnetic TTF radical as a dopant in an as-synthesized COF was found by a weak signal on electron spin resonance (ESR) spectra. Different degrees of doping on varying spin-spin interactions with cumulative concentrations of TTF radicals also reflect the increased paramagnetic intensity of approximately 1 and 3 orders of magnitude after exposure of TCNQ and I_2, respectively.

5.4 N-Type Semiconducting 2D COFs

The first report about the electron transporting 2D COF developed by Jiang et al. highlights the importance of embedded π-electronic building blocks in the conducting properties of the frameworks (Fig. 5.7) [27]. In their established strategy, an electron-deficient π-electronic building block such as benzothiazole (BTDA) was embedded at the edges of a two-component tetragonal nickel(II) phthalocyanine COF (Fig. 5.6). Integration of BTDA units at the edge causes a drastic change in the carrier transport mode from the hole transporting with a phenyl unit at the edge to electron-conducting frameworks. Co-condensation of MPc with 1,4-benzothiazole-diboronic acid (BTDADA) under solvothermal conditions leads to the formation of a crystalline belt-shaped 2D NiPc-BTDA COF that adopts an AA-type stacking arrangement with a tetragonal pore of diameter 2.2 nm and a surface area of 877 cm/g. Due to its preferable eclipsed stacking, it exhibits broad and enhanced absorbance up to 1000 nm, which motivated investigations of its photoconducting behavior. Upon irradiation with visible light (>400 nm), it exhibits an enhanced photocurrent from 250 μA to 15 μA at a 1 V bias, which runs down after many sequences. The high sensitivity toward NIR photons through a panchromatic response was investigated by using different band pass filters upon selective excitation with different wavelengths. FP-TRMC methods were utilized to investigate the intrinsic charge-carrier mobility under atmospheres of O_2, SF_6, and Ar after excitation with a 355 nm pulsed laser. The profile shows a double exponential decay curve, fitting for electrons and holes, under an Ar atmosphere and gives a $\phi\Sigma\mu$ value of 5.8×10^{-4} cm^2 V^{-1} s^{-1}. The decay profile responsible for electrons was augmented considerably upon measurement under SF_6 and O_2 atmospheres, which are well known as electron quenchers. In contrast, the decay profile responsible for holes was unaltered with respect to the measurement atmosphere. These results revealed a 2D NiPc-BTDA COF to be an electron transporting framework with electron mobility as high as 0.6 $cm^2\ V^{-1}\ s^{-1}$, which is outstanding.

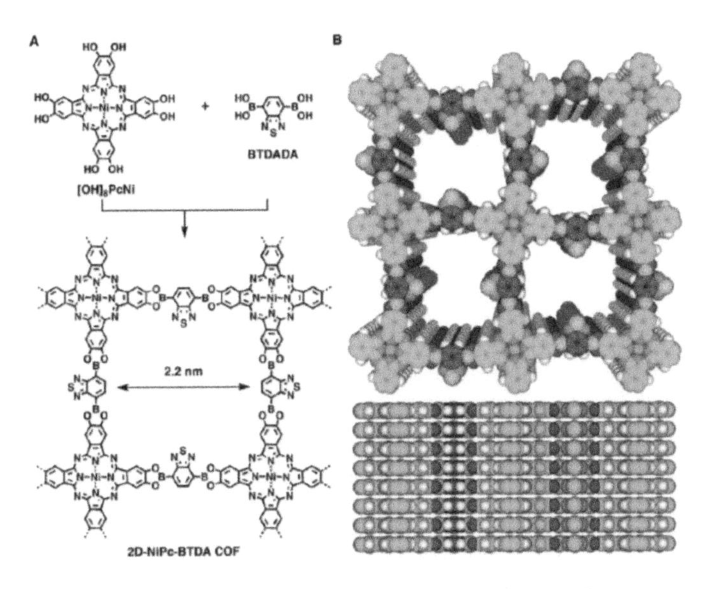

Figure 5.7 (A) Schematic representation of the synthesis of 2D NiPc-BTDA COF with metallophthalocyanine at the vertices and BTDA at the edges of the tetragonal framework and (B) top and side views of a graphical representation of a 2 × 2 tetragonal grid showing the eclipsed stacking of 2D polymer sheets. (Pc: sky blue; BTDA: violet; Ni: green; N: blue; S: yellow; O: red; B: orange; H: white). Reprinted with permission from Ref. [27]. Copyright (2011) American Chemical Society.

The same group reported one more n-type semiconducting 2D COF (CuP COF) with copper porphyrin units at the nodes of a tetragonal mesoporous framework surrounded by phenyl units at the edge (Fig. 5.4) [20]. The quantitative evolution of intrinsic electron carrier mobility was obtained by using the FP-TRMC technique in an Ar atmosphere. Upon laser flash, the rise and decay profiles of the FP-TRMC signal gave a maximum value of 1.16×10^{-4} $cm^2\ V^{-1}\ s^{-1}$ ($\phi\Sigma\mu$) for the CuP COF. The maximum quantum yield (ϕ) of photocarrier generation of the Cu COF was 6×10^{-4}, which led to the minimum electron mobility of 0.19 $cm^2 V^{-1} s^{-1}$. This distinct conducting nature of the CuP COF was also reflected in the photoconductivity, with a generated photocurrent of 0.6 nA upon irradiation with visible light. Thus, the central metal ion in this kind of porphyrin-based COF governs the type of charge-carrier motion within the frameworks. The eclipsed stacking of this COF offers the formation of two channels for carrier motion, that is macrocycle and metal channels. The presence of a copper ion at the core of the porphyrin macrocycle leads to the formation of a ligand-to-metal CT complex, which reduces the significant electron density on the porphyrin wall. Consequently, the metal-on-metal ordering on a CuP COF favors electron transport along the channel with high carrier mobility.

5.5 Ambipolar Semiconducting 2D COFs

As to the development of organic electronics, in particular for efficient solar cells with control over change dynamics, the macroscopic crystallization of the donor and acceptor moieties of the separated domains of donor and acceptor phases with control over their morphology is one of the ongoing challenges. In common materials, donor and acceptor moieties tend to stack one on top of the other to form a disordered assembly, where change carriers are trapped and readily annihilated through rapid combination. Such hurdles could be easily overcome by embedding a donor and an acceptor moiety into a 2D COF framework, which favors the production of periodic and unidirectional donor-on-donor and acceptor-on-acceptor columnar arrays with an ambipolar semiconducting behavior. To achieve the ambipolar semiconducting nature in COFs, various research groups either used postsynthesis functionalization or used donor and acceptor monomers directly in the synthesis.

At first, Jiang et al. demonstrated that donor and acceptor molecules can be embedded in a 2D COF framework, where self-sorting and co-crystallization between them leads to the formation of vertical unidirectional donor-on-donor and acceptor-on-acceptor columnar arrays [28]. These crystalline self-sorting and bicontinuous segregation alignments offer periodic independent pathways that allow ambipolar electron and hole conductions via acceptor-on-acceptor and donor-on-donor columns. Solvothermal condensation of HHTP as donor and BTDADA as acceptor leads to the formation of a donor-acceptor COF as a crystalline orange belt (Fig. 5.8a). Different characterization details exposed at the macroscopic level of the AA stacking mode shape the covalent 2D sheets into a framework with a pore diameter of 28 Å and a surface area of 2021 m^2/g. The distinct signature in the electronic absorption spectra at 425 nm with a decreasing HOMO-LUMO gap from 1.66 eV for the monolayer to 1.49 eV for the AA stacking forms indicates the charge-transporting ability in the stacked structure. FP-TRMC techniques in an Ar atmosphere were used to measure hole mobilities ($\phi\Sigma\mu$), including both electrons and holes, while a SF6 atmosphere was employed for the measurement of hole mobilities (μ_h). Accordingly, the hole and electron mobilities of a 2D donor-acceptor COF are 0.01 and 0.04 $cm^2\ V^{-1}\ s^{-1}$, respectively. A significant increase in current from 0.8 pA to 10.1 nA with a linear *I-V* curve was observed upon irradiation with visible light, which can be switched on and off many times without deterioration. They further highlight the choice of an appropriate donor-acceptor pair to explain the role of the lattice structure in CT and separation in donor-acceptor COFs with a pore size as large as 5.3 nm [29]. By exploiting triphenylene as a common vertex and two different diimides as edge units, Jiang et al. achieved two hexagonal triphenylene-diimide COFs (D_{TP}-A_{NDI} COF and D_{TP}-A_{PyDI} COF) with a high surface area and crystallinity (Fig. 5.8b). In comparison with individual counterparts, steady-state absorption spectra of the D_{TP}-A_{PyDI} COF showed a clear CT band, whereas no such band was observed for the D_{TP}-A_{NDI} COF. This implies that, independent of their lattice structures, the D_{TP}-A_{PyDI} COF forms a through-bond CT complex, whereas in the D_{TP}-A_{NDI} COF the donor and acceptor columns are independent at the ground state, giving rise to a neutral system ready for charge separation. Time-resolved ESR (TR-ESR) spectroscopy uncovered a rapid increase in intensity (up to 2.5 μs) for the D_{TP}-A_{NDI} COF as a result of charge separation. In sharp contrast, the D_{TP}-A_{PyDI} COF did not show any signals in the TR-ESR measurements under otherwise identical conditions. The fluorescence lifetime of the D_{TP}-A_{NDI} COF measured by using time-resolved fluorescence spectroscopy revealed a lifetime of 0.82 ns for an estimated charge-separation efficiency of 84%. Thus, the appropriate choice of the donor and acceptor pair is crucial to form the charge-separated state in COFs, a key process to achieve photoenergy conversion.

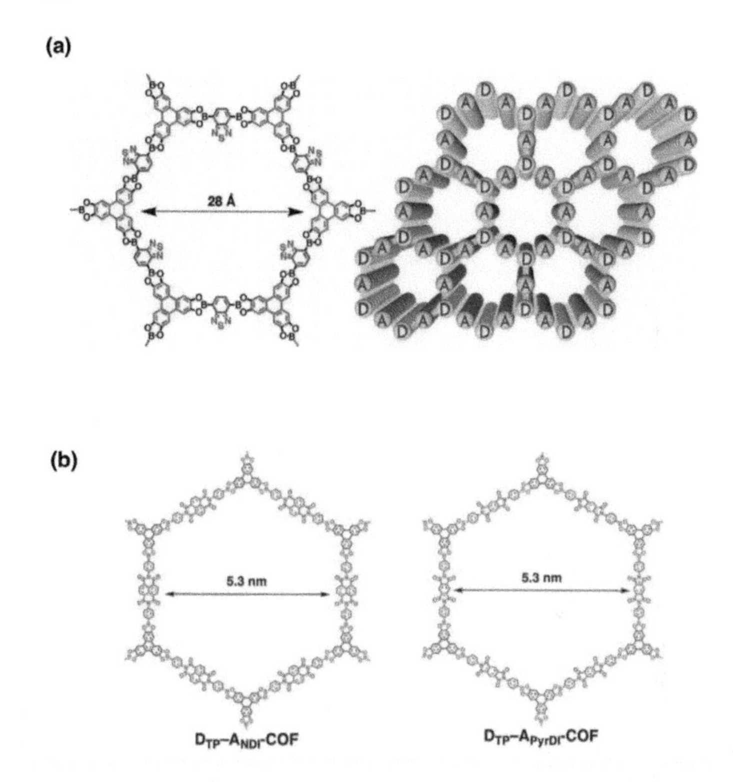

Figure 5.8 (a) Schematic representation and partial chemical structure of 2D donor-acceptor COF with self-sorting periodic donor-acceptor ordering and bicontinuous conducting channels. (b) Partial chemical structure of D_{TP}-A_{NDI} COF and DTP-APyDI COF. Reproduced with permission from Ref. [28]. Copyright (2012), John Wiley and Sons. Republished with permission of Royal Society of Chemistry, from Ref. [29], copyright (2013); permission conveyed through Copyright Clearance Center, Inc.

Bein et al. reported a porphyrin- and triphenylene-containing T_P-P COF featuring ordered columns of donor and acceptor moieties within its framework that promote paths for exciton and charge-carrier migration upon photoexcitation of either building block (Fig. 5.9) [30]. Photoinduced absorption spectroscopy measured on the T_P-P COF film grown on fused silica substrates exhibits two characteristic bands centered at 700 and 960 nm, like the sum of the

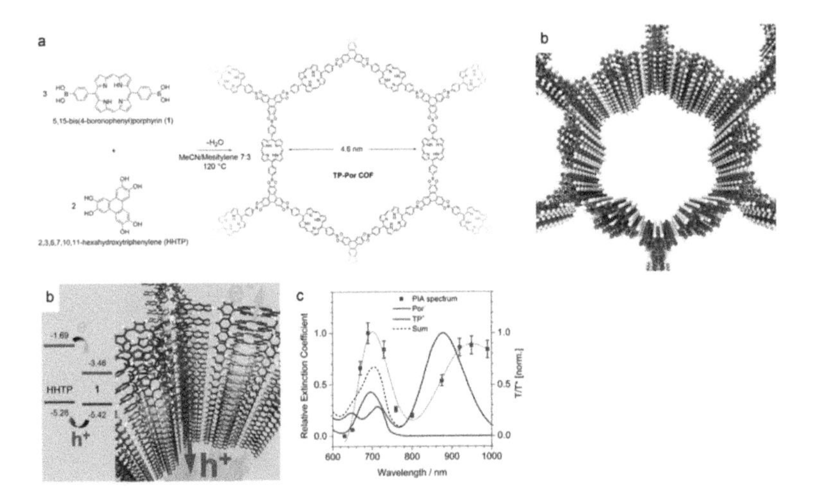

Figure 5.9 (a) Partial chemical structure of T_P-P COF with an illustration of the T_P-P COF highlighting the alternating columns of triphenylene (red) and porphyrin (blue) subunits. (b) Frontier orbital energy diagrams of the two COF subunits measured by DPV in solution and a schematic illustration of the photoinduced charge transfer. (c) Photoinduced absorption spectroscopy spectrum of the T_P-P COF film after excitation at 470 nm (blue squares; the blue line) together with the radical ion absorption spectra a 1:1 ratio of the two species. After photoexcitation, the TP-P COF film shows two absorption bands in the range of the free radical ion absorption, indicating electron transfer from the donor to the acceptor moiety within the network. Reprinted with permission from Ref. [30]. Copyright (2014) American Chemical Society <https://pubs.acs.org/doi/10.1021/ja509551m>. Further permissions related to the material excerpted should be directed to the ACS.

spectra of electron acceptor (P^-) and donor (TP^+) free radical ions (Fig. 5.9c). The energy-level diagram calculated from differential pulse voltammetry (DPV) techniques further supports this claim and indicates the possibility of efficient electron transfer within the framework (Fig. 5.9b). Vertically oriented thin 50 nm COF film sandwiched between ITO/MoO_x and ZnO/Al electrodes revealed the short-circuit current about 30 times higher than that of a reference device based on a randomly intermixed blend of the two building blocks. Upon illuminating the device with simulated solar light, the external quantum efficiency could be boosted to more than 30% at 350 nm and well above 10% up to 450 nm under a reverse bias. Quantum efficiency measurements in the presence of an external collection field show the potential of this novel device concept, provided that recombination losses can be minimized from further improvements in the electron and hole transport properties of these materials. Jiang and coworkers also highlighted a 2D porphyrin COF (ZnP COF), where the porphyrin moiety coordinated with zinc ion revealed an ambipolar conducting nature, as shown in Fig. 5.4 [20]. Intrinsic charge-carrier mobility of the ZnP COF measured by FP-TRMC methods shows comparable carrier mobilities for both holes and electrons ($\phi\mu_e$ = 5.4 × 10^{-5} cm^2 V^{-1} s^{-1}, $\phi\mu_h$ = 3.36 × 10^{-5} cm^2 V^{-1} s^{-1}). Transient absorption spectroscopic measurements exhibited the maximum quantum yield for photocarrier generation (ϕ) to be 1.9 × 10^{-3}, and the μ_e and μ_h values were 0.016 and 0.032 cm^2 V^{-1} s^{-1}, respectively. Photoconductivity measurements upon irradiation with visible light generated a photocurrent of approximately 26.8 nA with a large on-off ratio (5 × 10^4). Thus, an ambipolar conducting nature with a balanced hole and electron transporting ability drives the production of a large photocurrent as well as a high on-off ratio, as compared to unbalanced carrier transport.

Expanding on the intracrystalline donor-acceptor theme, Jiang et al. further elaborated their search on COF CT and separation, with a series of 2D COFs embedded with different metallophthalocyanates and diimides as electron-donating and -accepting building blocks (Fig. 5.10) [31, 32]. A series of COFs, namely D_{MPc}-A_{DI} COFs, were prepared by using solvothermal reactions, and it was shown that polycondensation is widely applicable for various MPcs, including copper, nickel, and zinc, with diimide and perylene diimide. Experimental and theoretical structural parameters were studied in all these COF frameworks, and it was found that MPc donor and diimide acceptor units are exactly linked and interfaced to form ordered segregated yet bicontinuous periodic π arrays.

All these COFs have light absorption over a wide range of the visible spectrum and into the NIR up to 1100 nm. The absence of the absorption characteristic after 1500 nm for D_{MPc}-A_{DI} COFs ruled out the possibility of ground-state interactions between MPc and DI to form a CT complex and support the existence of independent π columns. This type of structural feature offers two classes of π columns—four proximate directly linked central donor pairs and eight remote pairs without direct linkage—with different spatial center-to-center distances (<2 nm for the proximate pair and >4 nm for the remote pair). Thus, in such a framework four available equivalent acceptors are stacked in proximate columns for accepting the electron from an excited donor unit. Furthermore, the discrete center-to-center distances on D_{MPc}-A_{DI} COFs allow independent pathways for exciton migration and hole and electron transport along with fine screening to achieve long-lived charge separation. Time-resolved spectroscopy measured a dispersed solution of D_{MPc}-A_{DI} COFs in benzonitrile at room temperature upon irradiation of singlet spin and charge migration following the charge separation.

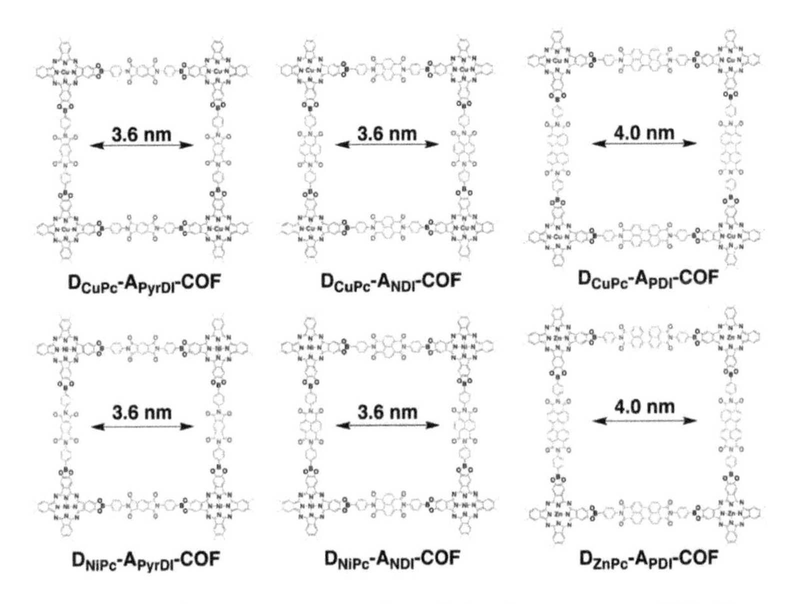

Figure 5.10 Schematic representation of the donor-acceptor COFs (D_{MPc}-A_{DI} COFs) with covalently linked phthalocyanine diimide (MPc-DI) structures. Reprinted with permission from Ref. [31]. Copyright (2015) American Chemical Society.

The lifetime values for charge separation were relatively longer on D_{CuPc}-A_{DI} COFs compared to those on D_{NiPc}-A_{DI} COFs and D_{ZnPc}-A_{DI} COFs. The higher oxidation states of CuPc compared to NiPc and ZnPc may cause charge recombination to occur by providing much higher larger driving forces for back electron transfer, responsible for this trend. A TR-ESR signal up to 1.35 μs was observed for D_{ZnPc}-A_{PDI} COFs, in sharp contrast to D_{ZnPc}-A_{PDI} COFs, where the paramagnetic nature of copper complicates the resolution of the spectral profile. In addition, the DPV measurements with a dispersion of D_{ZnPc}-A_{PDI} COFs with respect to a Ag/$AgNO_3$ reference electrode showed oxidation of the ZnPc at 0.42 eV and a reduction of the naphthalene diimide (NDI) at 0.52 eV, which the authors interpreted as a large exothermic driving force for electron transfer. Thus, the structural feature in these donor-acceptor COFs and resulting insights into the mechanistic aspects of charge separation make it possible to extract holes and electrons for electric current production as well as constitute a basis for developing COFs for photovoltaic applications.

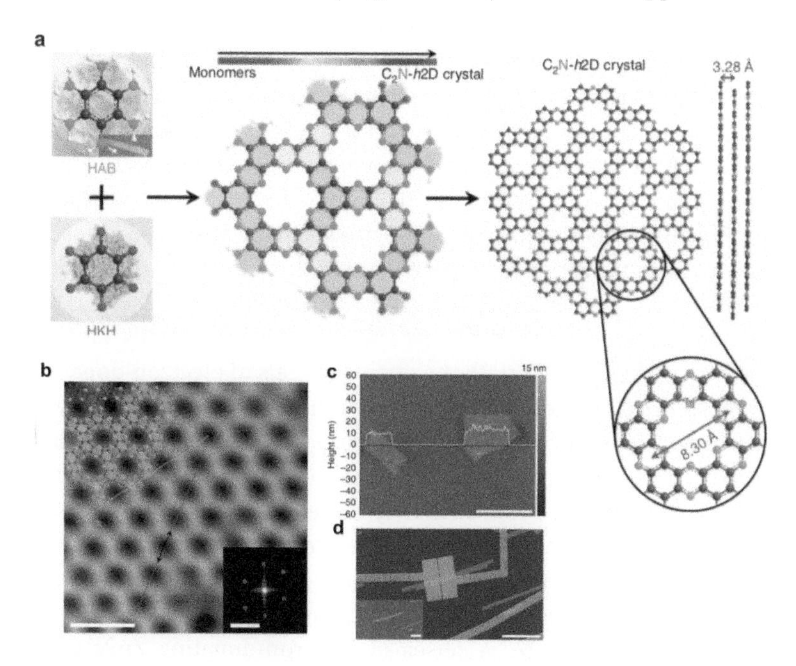

Figure 5.11 (a) Schematic representation of the reaction between HAB and HKH to provide the C_2H-h_2D crystal and their partial chemical structures. (b) An atomic force microscopy (AFM) image of the C_2H-h_2D crystal on Cu (111). (c) AFM image of the C_2H-h_2D crystal with height profile along the cyan-blue line (scale bar: 7 μm). (d) Optical microscopy image of a C_2H-h_2D crystal FET prepared on a SiO_3 (300 nm)/n^{++} Si wafer. The inset is an optical microscopy image taken before the deposition of Au electrodes on the crystal. Reproduced from Ref. [33] under a Creative Commons Attribution 4.0 International Licence (https://creativecommons.org/licenses/by/4.0/).

Recently, Baek and coworkers reported a 2D-layered network structure consisting of evenly distributed periodic holes and nitrogen atoms with a C_2N stoichiometry in its basal plane called "nitro-generated holey graphene" (C_2H-h_2D), as shown in Fig.

5.11 [33]. The wet chemical reaction between hexaaminobenzene trihydrochloride (HAB) and hexaketocyclohexane octahydrate (HKH) in *N*-methyl-2-pyrrolidone in the presence of sulfonic acid (catalytic amount) leads to the formation of a holey graphene framework, where phenyl rings are bridged by pyrazine rings. Density functional theory (DFT) calculations revealed a remarkably high potential energy (89.7 kcal U mol^{-1}) gain by the aromatization drive in this polycondensation reaction, which favors the formation of a 2D network structure. Further, the magnitude of the bandgap and the existence of flat bands near the Fermi levels support that the C_2H-h_2D as the active layer with a thickness of 8 nm, showing ambipolar charge transport with an electron mobility of 13.5 cm^2 V^{-1} s^{-1} and a hole mobility of 20.6 cm^2 V^{-1} s^{-1}. The unintentional doping effects by the trapped impurities and/or absorbed gases on the open framework are responsible for this behavior and further indicate that the electronic properties are tunable.

Jiang et al. further developed a strategy to spatially confine electron acceptor COFs with the open channels of electron-donating frameworks by exploring covalent click chemistry to achieve donor-acceptor COFs with an ambipolar conducting nature [34]. In such systems photoinduced electron transfer and charge separation, along with carrier concentration efficiency, largely depend on the acceptor content in the open pore. They developed a three-component topological design in conjunction with click chemistry for adapting an open-lattice COF into ordered donor-acceptor arrays with a photoelectric structure. A series of electron-donating ZnPc COFs were synthesized by three-component solvothermal condensation with Zn-phthalocyanine as the vertex and varying the molar ratio of BDBA and 2,5-bis(azidomethyl)-1,4-phenylenediboronic acid as edges (Fig. 5.12). The click reaction between the azide moieties of the channel wall with the alkyne derivatives of fullerene leads to channel confinements with the lattice altering into an ordered donor-acceptor system. A significant increase on the absorption band at 451 nm responsible for the fullerene moiety observed with increasing molar percentage of the azide unit on ZnPc COFs further supports this strategy. Upon the laser flash at 680 nm, the electron-donating ZnPc COF revealed a broad TR-ESR signal generated from the excited triplet states without the generation of charge-separated states. A drastic change of this on integration of electron-accepting fullerene molecules into the channel walls indicates light irradiation trigger electron transfer from the cation radical $ZnPc^{•+}$ to the anion radical $C_{60}^{•-}$ and delocalization of the radical cations and anions in the ZnPc and C_{60} columns. A significant increase in the signal intensity with increasing fullerene contents indicates that the radical species are increased in the framework, which increases the efficiency of the photoinduced electron transfer and charge separation.

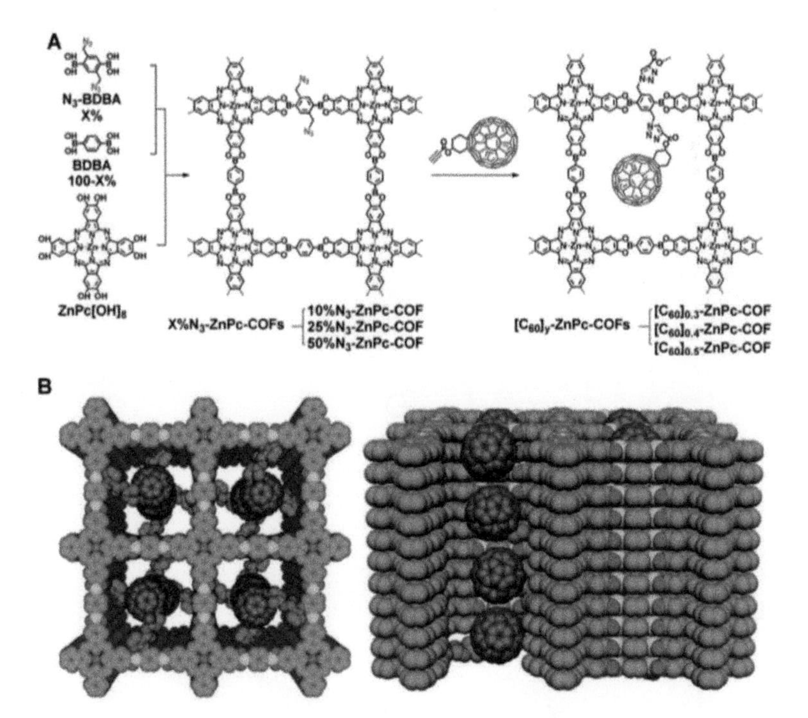

Figure 5.12 (A) Partial chemical structure of ZnPc COFs and the reaction to integrate a C_{60} molecule on the channel wall. (B) Schematic representation of top and side views of ZnPc COFs with C_{60} integrated on the channel walls. Reprinted with permission from Ref. [34]. Copyright (2014) American Chemical Society.

As to an alternative strategy, two groups researched a supramolecular approach by spatially confining fullerene molecules as electron acceptors within the open channels of electron-donating frameworks, for converting the open lattice of COFs into donor-acceptor photoelectric structures. Knochel et al. reported a COF-based photovoltaic device where light-induced charge separation and collection is clearly feasible with an electron donor in the walls and the complementary electron acceptor in ordered periodic channels (Fig. 5.13) [35]. Under solvothermal conditions, the condensation of thieno[3,2-b]thiophene-2,5-diyldiboronic acid with HHTP leads to the formation of thienothiophene-based TT COF with a surface area of 1810 m^2/g and a pore diameter of 3 nm. This open framework allows the uptake of electron acceptor [6,6]-phenyl-C_{60}-butyric acid methyl ester (PCBM) and favors the formation of a periodic ordered donor-acceptor network. Spectroscopic results containing

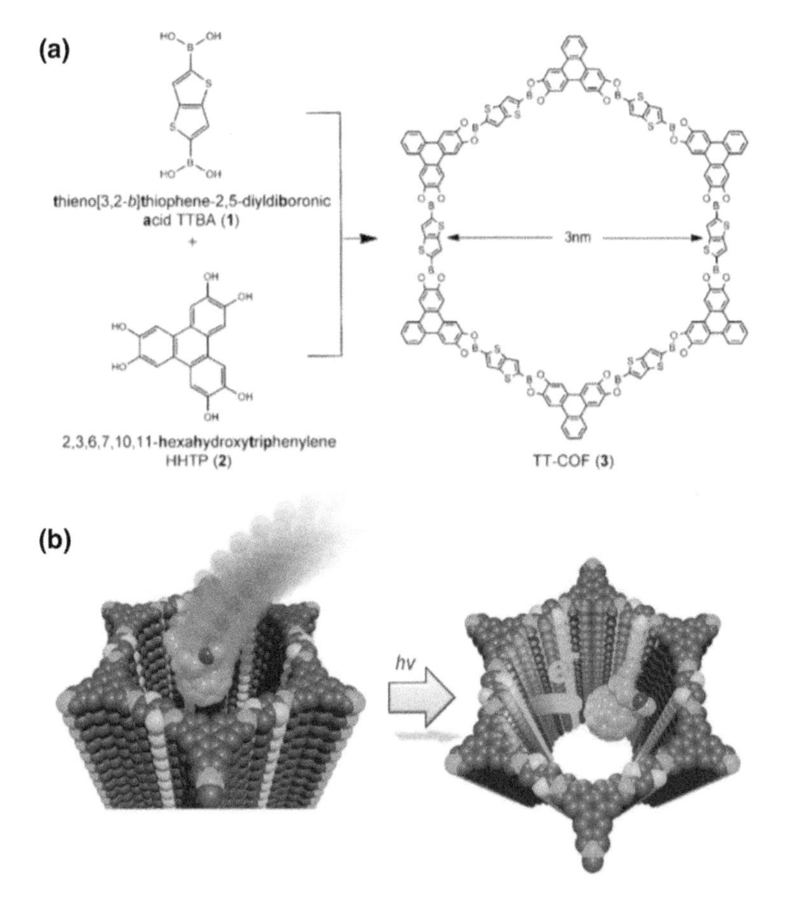

Figure 5.13 (a) Chemical structure of TT COF. (b) Schematic representation of the host-guest complex and a PCBM molecule to scale (C: gray; O: red; B: green; S: yellow). Reproduced from Ref. [35]. Copyright (2013), John Wiley and Sons.

absorption, emission, and time-resolved data demonstrate light-induced efficient CT within the lifetime of the excitons from the photoconductive TT COF donor network to the encapsulated PCBM phase thickness into a ITO/TT-COF:PCBM/Al device structure, which reveals a linear *I-V* profile with a power conversion efficiency of 0.053% under illumination with simulated solar light. The maximum external quantum efficiency of 3.4% at 405 nm matches well with the absorption characteristics of the COF film, whereas the tail at wavelengths above 700 nm indicates that the characteristic feature of PCBM with both interpenetrating networks is photoelectrically active.

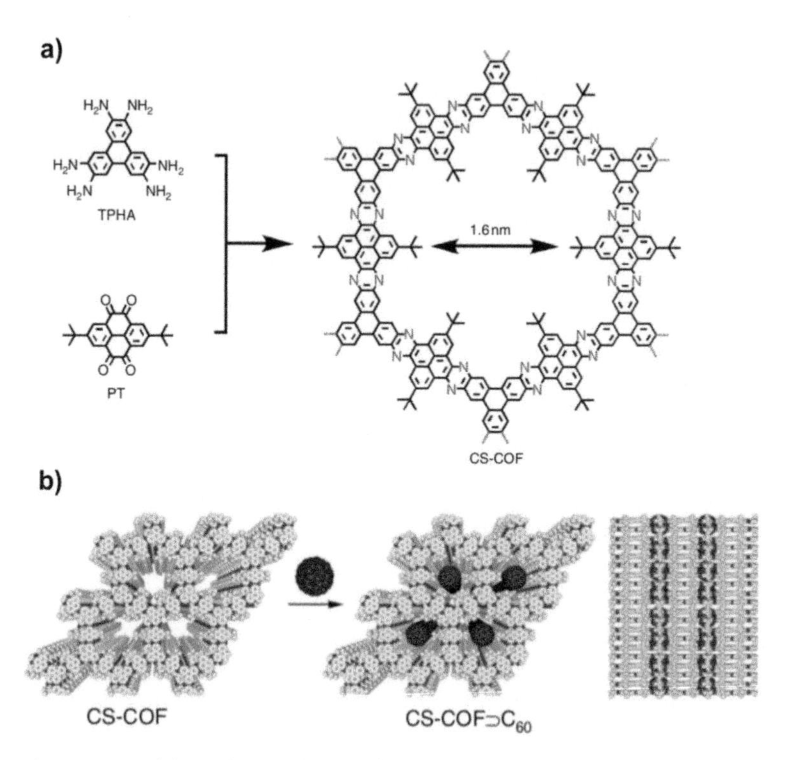

Figure 5.14 (a) Synthetic scheme of a CS COF. (b) Schematic representation of the synthesis of CS-COF-C_{60} by sublimed crystallization of fullerene in the open 1D channels (white: carbon; red: nitrogen; purple: fullerene). Reprinted by permission from Springer Nature Customer Service Centre GmbH: Springer Nature, *Nature Communications*, Ref. [36], copyright (2013).

More recently, Jiang et al. developed the synthesis of a crystalline phenazine-linked CS COF with an ordered stable structure and delocalized π clouds by topological ring fusion of triphenylene hexamine and *tert*-butylpyrene tetraone as the monomers (Fig. 5.14) [36]. The FP-TRMC method showed exceptional mobility of 4.2 $cm^2 V^{-1} s^{-1}$, suggesting that the CS COF is a high-rate hole-conducting framework, which is ranked among the top-class hole transporting organic semiconductors. Furthermore, the micropores of the CS COF allow complementary functionalization; physically filling the pore with fullerene molecules converts it to a bicontinuous order donor-acceptor system (CS-COF-C_{60}). Slipped AA stacking with bulky *tert*-butyl groups on the channel walls tolerates the accommodation of only one C_{60} molecule in the pore. To explore the possibility of photoenergy conversion, PCBM was used as a glue to fabricate 1 × 1 cm^2 ITO/active layer/Al cells using spin-coated CS-COF-C_{60} films with an adjustable thickness of 100 nm. The sandwiched devices with CS-COF-C_{60} as the photoconductive layer display a rapid response to light irradiation with an on-off ratio of up to 5.9 × 10^7 and a power conversion efficiency of 0.9%.

5.6 Lithium-Ion Batteries Using 2D COFs as Electrodes

Now, Li-ion batteries (LIBs) have attracted a wide range of interest because they can directly store energy, have great success in portable electronic products, and are a clean energy source that is expected to accelerate the development of electric vehicles [37, 38]. However, most commercially available LIBs are transition metal oxide cathodes, which will place a heavy burden on the ecological environment. For sustainable development, the battery cathode should preferably be made of highly efficient and environmentally friendly materials. For example, Pang et al. used a sulfur cathode—for example, in lithium-sulfur (Li-S) batteries—to take the place of conventional insertion cathode materials, or used organic cathode materials instead of transition metal oxide cathodes because they do not contain transition metal species [39, 40]. Porous COF materials possess high surface areas, and current efforts toward utilizing porous organic polymers in energy storage applications mainly focus on capacity improvement by the formation of electrochemical double layers, and no solid-state redox reactions take place within the framework itself. In this case, COFs as an electrode material in energy storage have attracted attention [41]. The structural advantages of 2D COFs, with ultrahigh surface areas, tunable pore sizes and shapes, and the ability to exhibit nanoscale effects on the pore wall, offer advantages for application of these kinds of materials for energy conversion and storage devices. This section on LIBs focuses on the application of 2D COF materials in battery anodes and cathodes, demonstrating the superior properties of this material in terms of electrical energy storage and being an environmentally friendly alternative.

5.6.1 Battery Cathode Application

Li-S batteries have relatively high theoretical energy density and gravimetric capacity and low production cost. They are considered to be promising energy storage devices, have received extensive attention, and are worthy of vigorous development [42, 43]. However, Li-S batteries possess the following problems in practical applications [44–47]. In the course of repeated charge and discharge cycles, a succession of structural and morphological alternations can occur in cyclo-S_8. This issue leads to the formation of soluble lithium polysulfides Li_2S_x and insoluble Li_2S_2/Li_2S. The polysulfides will dissolve in the electrolyte during the charging process of the battery, may shuttle back and forth between the anode and the cathode, and will have a side reaction with the anode and a reoxidation reaction with the cathode. These possible reactions will result in poor battery efficiency and stability [41, 43, 47].

Two-dimensional COFs are ordered structures with rigid covalent bond networks and pores. When applied to Li-S batteries, they may be conducive to the transport of electrolyte ions. In addition, the conductivity of sulfur can be increased by using the conductivity of 2D COFs. Wang et al. utilized CTF-1 (pore size of 1.23 nm) as a new host material for the insertion of sulfur to capture soluble polysulfides and improve the stability of the battery [48]. Figure 5.15A shows a sulfur-containing COF obtained by the melting and diffusion methods. The main experimental object of this experiment was CTF-1/S@155°C, which was obtained by heating a mixture of CFT-1 and sulfur (3:2) at 155°C for 15 h. Simple physical mixing of CTF-1 and sulfur, a mixed product labeled CTF-1/S@RT, was used for easy comparison [49, 50].

The XRD patterns of sulfur, CTF-1, and CTF-1/S@155°C were examined, as shown in Fig. 5.15B. Unlike sulfur and CTF-1, which have a number of strong peaks, CTF-1/S@155°C showed featureless and weak bands. CTF-1/S@155°C disappears on XRD patterns and the elements that make up CTF-1 are lightweight. This may indicate that sulfur is well dispersed within CTF-1 at the nanoscale and subnanometer levels. In addition, Pang et al. observed that the specific surface area (789–1.6 m^2/g) and pore volume (0.3–0.0036 cm^3/g) decrease sharply. This further confirmed that the pores of CTF-1 have been filled with sulfur and that sulfur occupies almost all of the pores [50].

The discharge curve of a typical CTF-1/S@155°C cathode (0.1 C, 1.1–3 V) is shown in Fig. 5.15C. The stage potential profile distribution is generated at the cathode of CTF-1/S@155°C. A shorter plateau is in a higher potential (2.3 V), which is consistent with the first reduction of sulfur to lithium polysulfides ($Li_2S\mathit{n}$, $2 < n < 8$), and the longer plateau is at a lower potential (~2.1 V), which is consistent with the solid lithium sulfides (Li_2S_2 and LiS) that are produced by further reduction of polysulfide. In Fig. 5.15C, the potential hysteresis between discharge and charge (0.2 V) shows the rapid kinetics of the transition between lithium polysulfide and lithium sulfide in the pores of the CTF-1 substrate. Obviously, the electrical touch of both embedded sulfur and the CTF wall facilitates these fast kinetics [50].

Figure 5.15D exhibits the cycling performance of the two materials acting as cathodes. In the first cycle, compared to the cathode discharge capacity for CTF-1/S@RT, CTF-1/S@155°C displayed higher values, indicating that the diffusion process is favored at high temperatures. Also, the S8 molecule may be strongly adsorbed in the nanopore of CTF-1 and the charge–discharge coulombic efficiency of the composite material is observed to be close to ~97%, as a result of which the soluble polysulfide intermediate can effectively inhibit the shuttle effect. After the third cycle, the capacity value becomes relatively stable, and after 20 cycles, the value is maintained at approximately 1000 mAh g^{-1}. It can be stated that the stable cycle performance of the CTF-1/S@155°C cathode is determined by the CTF-1 substrate and sulfur loading in CTF-1 pores can significantly increase the sulfur electrode cycle performance.

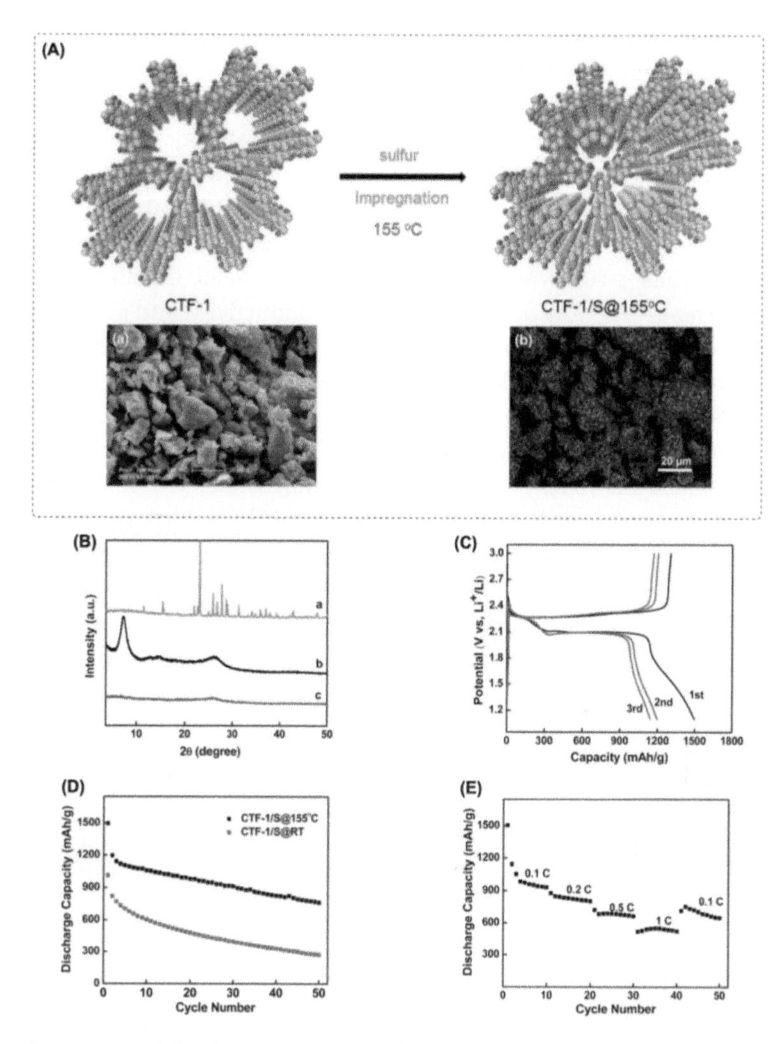

Figure 5.15 (A) Schematic diagram of the composite synthesis from CTF-1 by impregnation of molten sulfur (C: gray; N: blue; H: red; S: yellow) and SEM image of (a) the corresponding elemental mapping of sulfur (b) for the CTF-1/S@155°C composite. (B) Galvanostatic discharge and charge profiles of the CTF-1/S@155°C composite at a 0.1 C rate. (C) XRD patterns of sublimed sulfur (a), CTF-1 (b), and the CTF-1/S@155°C composite (c). (D) Cycling performance of the CTF-1/S@155°C composite and the CTF-1/S@RT composite at a 0.1 C rate. (E) Discharge capacity for the CTF-1/S@155°C composite at different rates. Republished with permission of Royal Society of Chemistry, from Ref. [48], copyright (2014); permission conveyed through Copyright Clearance Center, Inc.

Even if the current density increases, the discharge capacity does not drop rapidly, as shown in Fig. 5.15E. Although when the high current is high (1 C), the CTF-1/S@155°C cathode exhibits excellent rate performance, it can provide a reversible capacity of 541 mAh g^{-1}. More importantly, if the current density of 0.1 C is again applied, the capacity value will return to approximately 750 mAh g^{-1}. Apparently, due to the presence of CTF-1 substrates with ordered nanoporous structures, composite materials yield excellent rate performance, greatly improve the sulfur conductivity, and provide a tunnel for electrolyte penetration. This work offered possibilities for the development of 2D COFs in new energy-related applications.

With carbon nanotubes (CNTs) as conducting materials, Jiang et al. demonstrated a new method to employ the CNT–2D COF composite as an electrode [51]. Through an in situ polycondensation process, the redox-active COF (Fig. 5.16, D_{TP}-A_{NDI} COF), which contained NDI units, could be formed on the surface of CNT wires. Due to the redox properties of NDI units, the as-formed 2D COFs and D_{TP}-A_{NDI}-COF@CNTs composite were redox-active. From the XRD analysis of the composite, it was found that the 2D COFs formed ordered mesoporous channels and adopted an AA stacking mode. Using this composite as the cathode, the performance was investigated. After 100 cycles, the LIBs with D_{TP}-A_{NDI}-COF@CNT cathodes could present high energy-storage stability and nearly 100% coulombic efficiency. In the charge–discharge curves presented as shown in Fig. 5.16g, no obvious polarization could be observed as the profiles retained similar shapes. This indicated that both electrons and ions inside the batteries could be transported efficiently, which could be beneficial for the rapid charge–discharge processes. Despite the relatively low capacity, the battery showed good stability. When the current density increased to 12 C, the battery could still retain 85% of the capacity value at 2.4 C. At the rate of 2.4 C, the long-term stability was tested (Fig. 5.16i). Both capacity and coulombic efficiency were almost unchanged even after 700 cycles, indicating its excellent stability. The above examples imply that the pores inside 2D COFs could be used to trap energy-related species such as sulfur, polysulfides, and ions. What's more, when the 2D COFs are used in batteries, the advantage of possible conductivity against other π-conjugated materials could contribute to the final device performance.

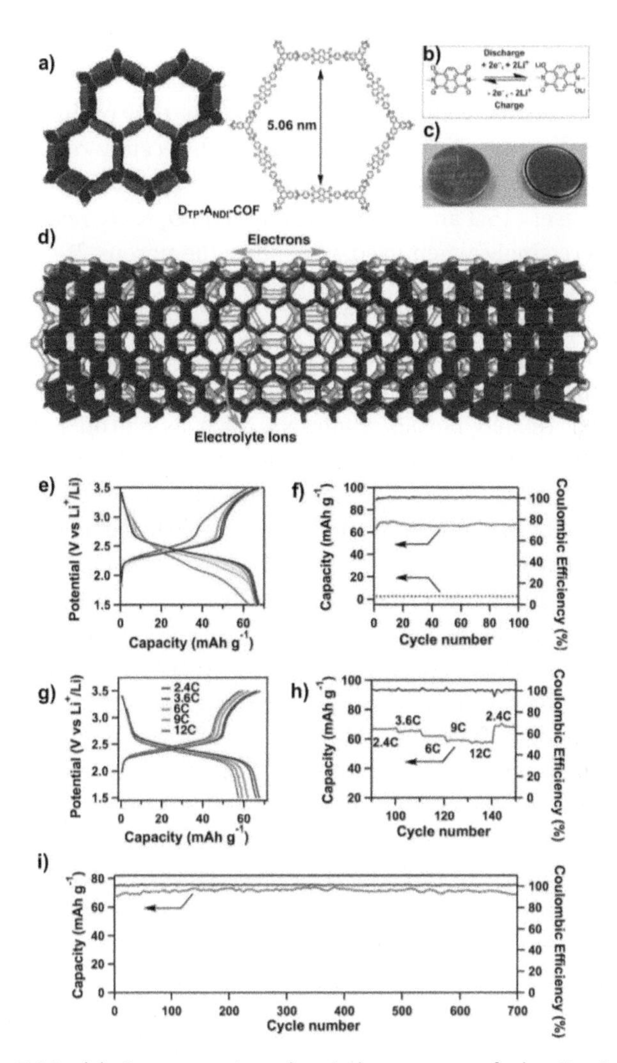

Figure 5.16 (a) Representative chemical structure of the D_{TP}-A_{NDI} COF composite. (b) Redox-active positions on the naphthalene diimide unit. (c) Pictures of the prepared coin-type lithium-sulfur batteries. (d) Schematic representation of the composite (gray color represents the CNTs). (e) Charge–discharge profiles of the composite cathodes (2.4 C; red is the 1st cycle, while blue is the 80th cycle). (f) Capacities of the composite cathodes and CNT (black dotted line) batteries and coulombic efficiency of the composite cathodes for 100 cycles. (g) Charge–discharge profiles of the composite cathodes at various charge–discharge rates. (h) Discharge capacity results of the composite cathodes at different rates. (i) Long-time stability result of the composite cathodes at the rate of 2.4 C. Reprinted by permission from Springer Nature Customer Service Centre GmbH: Springer Nature, *Scientific Reports*, Ref. [51], copyright (2015).

5.6.2 Battery Anode Application

Besides 2D COFs as cathode electrodes in LIBs, they could also serve as anode materials. For example, Zhao et al. recently reported two novel 2D crystalline COFs that could exhibit good selectivity for different gases (15:1 and 7:1 for H_2 to N_2 and CO_2 to N_2, respectively). More interestingly, when used as anodes, the porous 2D COFs also showed superior performance in Li-ion storage [52]. The average diameters of channels for N2 COF and N3 COF were measured to be 23 Å and 16 Å, respectively (Fig. 5.17). Owing to the porosity and high conductivity, the 2D COFs could exhibit better performance than other related porous materials, such as silicon and metal-organic frameworks (MOFs). In the charge–discharge curves presented in Figs. 5.18a and 5.18b, the 2D COF–based batteries didn't exhibit obvious polarization, which could be beneficial to both charge and discharge processes. For the batteries based on N2 COF (Fig. 5.18c), the charge capacity was 497 mA h g^{-1} at the rate of 5 C. The capacity could increase to 607 mA h g^{-1} when a current density of 2 C was adopted. For the N3-COF-based batteries (Fig. 5.18d), a much better performance was observed. The charge capacity values were maintained at ~520 mA h g^{-1} at the relatively high rate of 5 C. Furthermore, the capacity value of ~610 mA h g^{-1} could be recovered when the current density of 2 C was applied again. The cycling experiments (Fig. 5.18e,f) disclosed that the stabilities of both N2-COF- and N3-COF-based batteries were excellent.

Good conductivity is vital for electrode materials in LIBs with high performance. In the reported results, 2D COFs were mainly constructed through covalent connections based on several building blocks, such as boronate esters, borazine, enamine, imine, hydrazone, and triazine [53–57]. Generally, the conductivity of the as-formed COFs is poor compared to that of the conductive polymers. Porphyrin is a universal unit with unique optoelectronic properties, and it has also been incorporated into several 2D COFs. However, limited by the covalent linkage with boronic acids, its conductivity is not high enough for its application as an electrode material. With the purpose of further enhancing the conductivity of porphyrin-based 2D COFs, (4-thiophenephenyl)porphyrin (TThPP) was used as the building units for application as electrode materials (Fig. 5.19) [58] since thiophene is a universal unit in organic electronics, which could increase the planarity and electrical conductivity of the as-prepared materials.

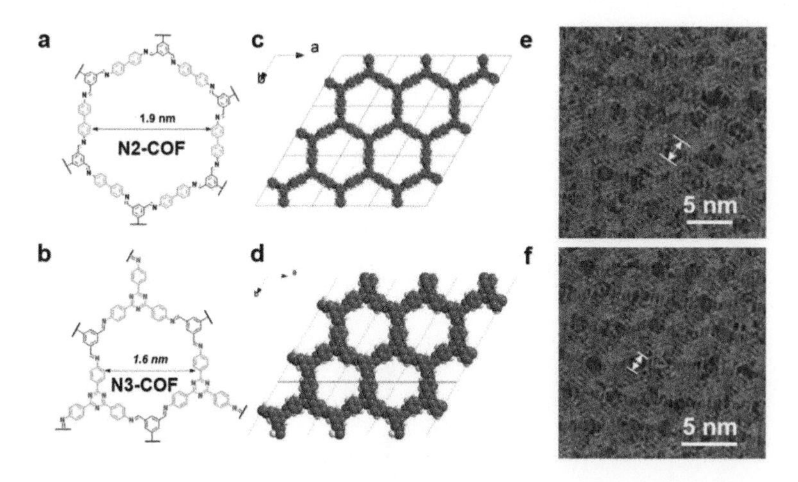

Figure 5.17 Structural presentation of the 2D COFs (a, b) and (c, d) the layered stacking mode (top view) based on theoretical DFT calculations (C: blue; N: red; H: gray-white). (e, f) STM images for the two 2D COFs. Republished with permission of Royal Society of Chemistry, from Ref. [52], copyright (2016); permission conveyed through Copyright Clearance Center, Inc.

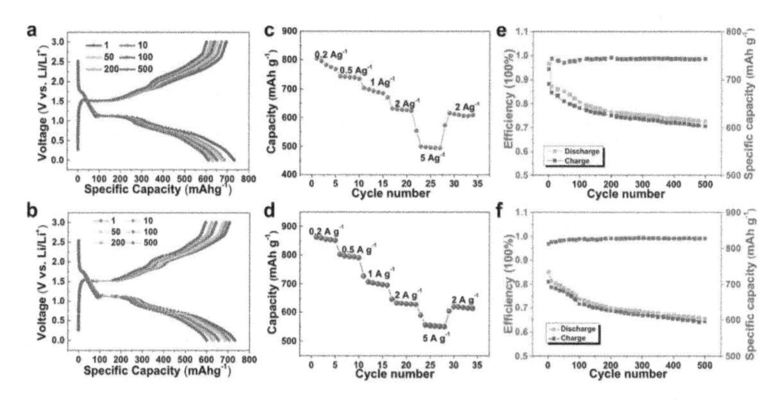

Figure 5.18 Charge–discharge curves of the 2D COFs (a) N2 COF and (b) N3 COF; current density of 1 Ag-1 was applied. Rate capability results at different current densities: (c) N2 COF and (d) N3 COF. Cycling performance of N2 COF (e) and N3 COF (f). The red profiles show the efficiency that was derived from the ratio of charge and discharge capacities. Republished with permission of Royal Society of Chemistry, from Ref. [52], copyright (2016); permission conveyed through Copyright Clearance Center, Inc.

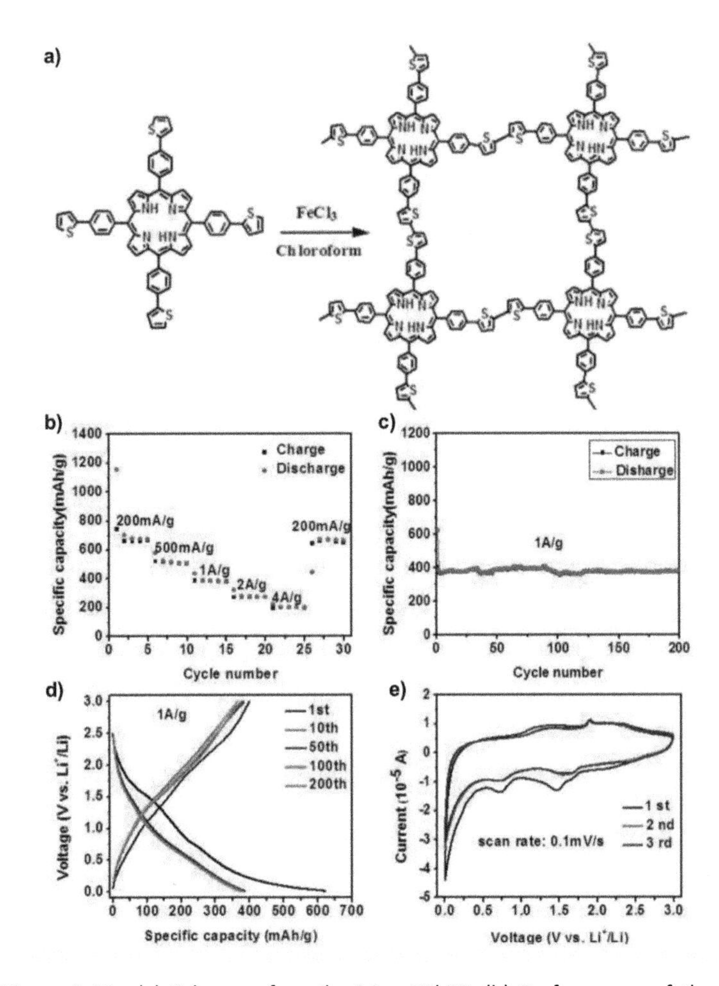

Figure 5.19 (a) Scheme of synthesizing TThPP. (b) Performance of the 2D COF–based anode; charge–discharge capacities at various rates. (c) Cycling experiment results at the rate of 1 C. (d) Charge–discharge curves of the anode from 1 to 200 cycles. (e) CV curves of the TThPP-based anode for 1 to 3 cycles. Reprinted with permission from Ref. [58]. Copyright (2016) American Chemical Society.

Using the chemical activity of thiophene units, the 2D COFs could be prepared via an oxidation polymerization reaction [59–61]. When the as-synthesized COFs were used as anode materials in batteries, the capacity value was 666 mA h g^{-1} at the current density of 0.2 C. At the rate of 1 C, a capacity value of 384 mA h g^{-1} could be observed.

When the rate changed from 5 to 0.2 C, the capacity value could be recovered to nearly 100% of the original capacity, indicating the stability of the batteries. The cycling experiment results are shown in Fig. 5.19b. After the charge–discharge process, the coulombic efficiency was calculated to be 99.3% while the discharge capacity of 384 mA h g^{-1} could be maintained even after 200 charge–discharge cycles.

Besides COFs, it is to be noted that Tarascon et al. pioneered the usage of MOFs as electrode materials in Li-S batteries [62]. The MOF MIL100(Cr) mixed with sulfur as the composite cathode could improve the cycling performance, even though the sulfur content was limited. Inspired by this, Xiao et al. demonstrated that Ni MOF ($Ni_6(BTB)_4(BP)_3$ (BTB: benzene-1,3,5-tribenzoate; BP: 4,4′-bipyridyl) could remarkably immobilize the sulfur and polysulfide species within the cathode structure [63]. As a host material, the Ni MOF/S composite-based batteries exhibited a discharge capacity of 689 mA h g^{-1} at 0.1 C. At a current density of 2 C (Table 5.2), the composite was capable of delivering a discharge capacity of 287 mA h g^{-1}. Li et al. studies several MOFs as host materials in Li-S batteries [64]. Results showed that the nanosized ZIF-8 and MIL-53 could serve as good host materials to produce a fast sulfur cathode with a long cycle life. Under a constant rate of 0.5 C, the maximum discharge capacities for S/HKUST-1, S/ZIF-8, and S/MIL-53 were 431, 738, and 793 mA h g^{-1}, respectively.

Use of graphene composites in LIBs has been extensively pursued for a long time. The $LiFePO_4$/graphene composite as a cathode electrode exhibited discharge capacities of 165 and 88 mA h g^{-1} at 0.1 C and 10 C, respectively [65]. Explored by Choi et al., vanadium pentoxide (V_2O_5) mixed with graphene as a cathode electrode showed a discharge capacity of 94.4 mA h g^{-1} at 10 C, with ultrahigh stability (100,000 cycles) [66]. Instead of cathode materials, as anode materials, they generally showed better performance [73]. Cheng et al. reported that composites of graphene nanosheets (GNSs) and Fe_3O_4 could exhibit 580 and 520 mA h g^{-1} at current densities of 0.7 and 1.75 C, respectively [67]. Furthermore, the analogue of borocarbonitrides could also show a capacity of 150 mA h g^{-1} at a current density of 1 C [69]. The inorganic counterparts MoS_2 and WS_2 could also be functionalized as electrode materials. For example, Cho et al. reported MoS_2 nanoplates that contained disordered graphene-like layers that could work as high-rate lithium battery anode materials. The capacity was 700 mA h g^{-1} even at the rate of 50 C [70].

Table 5.2 Performance of Li-ion batteries with representative 2D COFs and related materials

Category	Materials	Electrode	Capacity (mA h g^{-1})	Current density (C)	Ref.
2D COF	CTF-1/ S@155°C	Cathode	541	1	[50]
2D COF	Por-COF/S	Cathode	670	1	[42]
2D COF	DTP-ANDI-COF@CNT	Cathode	58	12	[51]
2D COF	N2 COF	Anode	497	5	[52]
2D COF	N3 COF	Anode	~520	5	
2D COF	TThPP-based COF	Anode	384	1	[58]
MOF	Ni MOF/S	Cathode	287	2	[63]
MOF	S/HKUST-1	Cathode	431	0.5	[64]
MOF	S/ZIF-8	Cathode	738	0.5	[64]
MOF	S/MIL-53	Cathode	793	0.5	[64]
Graphene	Graphene-wrapped $LiFePO_4$/C	Cathode	88	10	[65]
Graphene	V_2O_5/graphene	Cathode	94.4	10	[66]
Graphene	rGO/Fe_3O^4	Anode	520	1.75	[67]
Graphene	GNS/C_{60}	Anode	784	0.05	[68]
$B_xC_yN_z$	$B_xC_xN_z$ (CVD)	Anode	150	1	[69]
MoS_2	Nanosheets	Anode	700	50	[70]
MoS_2	TO/GS composite	Anode	199	1	[71]
WS_2	Composite	Anode	596	1.4	[72]

On performance comparison, it can be seen that 2D COF–based electrodes can show capacity comparable to that of other similar materials (NOFs, graphene, $B_xC_yN_z$, and MoS_2). 2D COFs are promising candidates as electrode materials in LIBs. However, how to achieve both high capacitance and long cycling performance is still the major bottleneck in using COFs as electrode materials for energy storage, probably due to the structural defects, low energy level matching, and insufficient device optimization during the charge–discharge process. From this viewpoint, further exploration to prepare new 2D COFs and gain ideas from other better-performing materials is highly desirable.

5.7 Summary and Perspective

This chapter introduced the recent progress in terms of some representative 2D COFs for energy-related applications, including conductivity study, semiconductor, lithium batteries, and capacitive storage. To fully understand the advantages and disadvantages of 2D COFs, a comparison study with other related materials (e.g., MOFs, graphene, borocarbonitrides, and MoS_2) was also conducted. Although research on 2D COFs has already attracted some attention, poor solubility is still the main obstacle for their usage beyond gas adsorption and storage (described in Chapter 3). In fact, according to literature results, energy-related applications of 2D COFs are still in their infancy. However, with more innovation, these unique 2D materials should have more potential applications. For example, in rechargeable lithium batteries, incorporation of sulfur into the pores of 2D COFs is a smart way to make them useful as electrode materials in energy-related applications. Also, utilizing the effective click reaction to introduce redox groups into conventional COFs would be a good way to provide 2D COFs with more chances for application in capacitive energy storage.

Finally, the true structure of the COF and the layer stacking sequence (interlaced or overlapping) remains controversial. Furthermore, the chemical stability of crystalline and ordered COF materials is poor, primarily because the formation reaction is limited by the reversible reaction of the formation of borate and imine double bonds. Therefore, another important aspect that needs to be explored is the development of a novel reaction and joining method to synthesize a class of COFs with long-range order and high chemical stability.

Although many challenges still exist, numerous possibilities are offered within this fascinating area and the authors of this book envision a bright future for the development of COF materials for application across fields, which will grow into an important research area with great potential.

References

1. Huang, N., Wang, P., Jiang, D. (2016). *Nat. Rev. Mater.*, **1**, 16068.
2. Dogru, M., Bein, T. (2014). *Chem. Commun.*, **50**, 5531–5546.
3. Feng, X., Ding, X., Jiang, D. (2012). *Chem. Soc. Rev.*, **41**, 6010–6022.
4. Mahmood, J., Lee, E. K., Jung, M., Shin, D., Choi, H.-J., Seo, J.-M., Jung, S.-M., Kim, D., Li, F., Lah, M. S., Park, N., Shin, H -J., Oh, J. H., Baek, J.-B. (2016). *Proc. Natl. Acad. Sci.*, **113**, 7414–7419.
5. Huang, N., Krishna, R., Jiang, D. (2015). *J. Am. Chem. Soc.*, **137**, 7079–7082.
6. Huang, N., Chen, X., Krishna, R., Jiang, D. (2015). *Angew. Chem. Int. Ed.*, **54**, 2986–2990.
7. Xu, H., Tao, S., Jiang, D. (2016). *Nat. Mater.*, **15**, 722–726.
8. Ma, H., Liu, B., Li, B., Zhang, L., Li, Y.-G., Tan, H.-Q., Zang, H.-Y., Zhu, G. (2016). *J. Am. Chem. Soc.*, **138**, 5897–5903.
9. Dalapati, S., Jin, E., Addicoat, M., Heine, T., Jiang, D. (2016). *J. Am. Chem. Soc.*, **138**, 5797–5800.
10. Ding, S.-Y., Dong, M., Wang, Y.-W., Chen, Y.-T., Wang, H.-Z., Su, C.-Y., Wang, W. (2016). *J. Am. Chem. Soc.*, **138**, 3031–3037.
11. Ostroverkhova, O. (2016). *Chem. Rev.*, **116**, 13279–13412.
12. Wan, S., Guo, J., Kim, J., Ihee, H., Jiang, D. (2008). *Angew. Chem. Int. Ed.*, **47**, 8826–8830.
13. Wan, S., Guo, J., Kim, J., Ihee, H., Jiang, D. (2009). *Angew. Chem. Int. Ed.*, **48**, 5439–5442.
14. Ding, X., Guo, J., Feng, X., Honsho Y., Guo, J., Seki, S., Maitarad, P., Saeki, A., Nagase, S., Jiang, D. (2011). *Angew. Chem. Int. Ed.*, **50**, 1289–1293.
15. Ding, X., Feng, X., Saeki, A., Seki S., Nagai, A., Jiang, D. (2012). *Chem. Commun.*, **48**, 8952–8954.
16. Liao M.-S., Scheiner, S. (2001). *J. Chem. Phys.*, **114**, 9780–9791;
17. Liao M.-S., Watts, J. D., Huang, M.-J., Gorun S. M., Kar, T., Scheiner, S. (2005). *J. Chem. Theory Comput.*, **1**, 1201–1210.

18. Dalapati, S., Addicoat, M., Jin, S., Sakurai, T., Gao, J., Xu, H., Irle, S., Seki, S., Jiang, D. (2015). *Nat. Commun.*, **6**, 7786.
19. Wan, S., Gµndara, F., Asano, A., Furukawa, H., Saeki, A., Dey, S. K., Liao, L., Ambrogio, M. W., Botros, Y. Y., Duan, X., Seki, S., Stoddart, J. F., Yaghi, O. M. (2011). *Chem. Mater.*, **23**, 4094–4097.
20. Feng, X., Liu, L., Honsho, Y., Saeki, A., Seki, S., Irle, S., Dong, Y., Nagai, A., Jiang, D. (2012). *Angew. Chem. Int. Ed.*, **51**, 2618–2622.
21. Segura, J. L., Martìn, N. (2001). *Angew. Chem. Int. Ed.*, **40**, 1372–1409.
22. Bertrand, G. H. V., Michaelis, V. K., Ong, T.-C., Griffin, R. G., Dincã, M. (2013). *Proc. Natl. Acad. Sci.*, **110**, 4923–4928.
23. Duhović, S., Dincã, M. (2015). *Chem. Mater.*, **27**, 5487–5490.
24. Jin, S., Sakurai, T., Kowalczyk, T., Dalapati, S., Xu, F., Wei, H., Chen, X., Gao, J., Seki, S., Irle, S., Jiang, D. (2014). *Chem. Eur. J.*, **20**, 14608–14613.
25. Ding, H., Li, Y., Hu, H., Sun, Y., Wang, J., Wang, C., Wang, C., Zhang, G., Wang, B., Xu, W., Zhang, D. (2014). *Chem. Eur. J.*, **20**, 14614–14618.
26. Cai, S.-L., Zhang, Y.-B., Pun, A. B., He, B., Yang, J., Toma, F. M., Sharp, I. D., Yaghi, O. M., Fan, J., Zheng, S.-R., Zhang, W.-G., Liu, Y. (2014). *Chem. Sci.*, **5**, 4693–4700.
27. Ding, X., Chen, L., Honsho, Y., Feng, X., Saengsawang, O., Guo, J., Saeki, A., Seki, S., Irle, S., Nagase, S., Parasuk, V., Jiang, D. (2011). *J. Am. Chem. Soc.*, **133**, 14510–14513.
28. Feng X., Chen, L., Honsho, Y., Saengsawang, O., Liu, L., Wang, L., Saeki, A., Irle, S., Seki, S., Dong, Y., Jiang, D. (2012). *Adv. Mater.*, **24**, 3026–3031.
29. Jin, S., Furukawa, K., Addicoat, M., Chen, L., Takahashi, S., Irle, S., Nakamura, T., Jiang, D. (2013). *Chem. Sci.*, **4**, 4505–4511.
30. Calik, M., Auras, F., Salonen, L. M., Bader, K., Grill, I., Handloser, M., Medina, D. D., Dogru, M., Lçbermann, F., Trauner, D., Hartschuh, A., Bein, T. (2014). *J. Am. Chem. Soc.*, **136**, 17802–17807.
31. Jin, S., Supur, M., Addicoat, M., Furukawa, K., Chen, L., Nakamura, T., Fukuzumi, S., Irle, S., Jiang, D. (2015). *J. Am. Chem. Soc.*, **137**, 7817–7827.
32. Jin, S., Ding, X., Feng, X., Supur, M., Furukawa, K., Takahashi, S., Addicoat, M., El-Khouly, M. E., Nakamura, T., Irle, S., Fukuzumi, S., Nagai, A., Jiang, D. (2013). *Angew. Chem. Int. Ed.*, **52**, 2017–2021.
33. Mahmood, J., Lee, E. K., Jung, M., Shin, D., Jeon, I.-Y., Jung, S.-M., Choi, H.-J., Seo, J.-M., Bae, S.-Y., Sohn, S.-D., Park, N., Oh, J. H., Shin, H.-J., Baek, J.-B. (2015). *Nat. Commun.*, **6**, 6486.
34. Chen, L., Furukawa, K., Gao, J., Nagai, A., Nakamura, T., Dong, Y., Jiang, D. (2014). *J. Am. Chem. Soc.*, **136**, 9806–9809.

35. Dogru, M., Handloser, M., Auras, F., Kunz, T., Medina, D., Hartschuh, A., Knochel, P., Bein, T. (2013). *Angew. Chem. Int. Ed.*, **52**, 2920–2924.

36. Guo, J., Xu, Y., Jin, S., Chen, L., Kaji, T., Honsho, Y., Addicoat, M. A., Kim, J., Saeki, A., Ihee, H., Seki, S., Irle, S., Hiramoto, M., Gao, J., Jiang, D. (2013). *Nat. Commun.*, **4**, 2736.

37. Xie, J., Gu, P., Zhang, Q. (2017). *ACS Energy Lett.*, 2, 1985–1996.

38. Xie, J., Wang, Z., Xu, Z. J., Zhang, Q. (2018). *Adv. Energy Mater.*, **8**, 1703509.

39. Zhan, X., Chen, Z., Zhang, Q. (2017). *J. Mater. Chem. A*, **5**, 14463.

40. Yang, X., Dong, B., Zhang, H., Ge, R., Gao, Y., Zhang, H. (2015). *RSC Adv.*, **5**, 86137.

41. Zhang, Y., Riduan, S. N., Wang, J. (2017). *Chem. Eur. J.*, **23**, 16419.

42. Liao, H., Wang, H., Ding, H., Meng, X., Xu, H., Wang, B., Ai, X., Wang, C. (2016). *J. Mater. Chem. A*, **4**, 7416–7421.

43. Yang, X., Dong, B., Zhang, H., Ge, R., Gao, Y., Zhang, H. (2015). *RSC Adv.*, **5**, 86137.

44. Li, N., Zheng, M., Lu, H., Hu, Z., Shen, C., Chang, X., Ji, G., Cao, J., Shi, Y. (2012). *Chem. Commun.*, **48**, 4106.

45. Zhao, Z., Wang, S., Liang, R., Li, Z., Shi, Z., Chen, G. (2014). *J. Mater. Chem. A*, **2**, 13509.

46. Gu, P.-Y., Zhao, Y., Xie, J., Binte Ali, N., Nie, L., Xu, Z. J., Zhang, Q. (2016). *ACS Appl. Mater. Interfaces*, **8**, 7464.

47. Ghazi, Z. A., Zhu, L., Wang, H., Naeem, A., Khattak, A. M., Liang, B., Khan, N. A., Wei, Z., Li, L., Tang, Z. (2016). *Adv. Energy Mater.*, **6**, 1601250.

48. Liao, H., Ding, H., Li, B., Ai, X., Wang, C. (2014). *J. Mater. Chem. A*, **2**, 8854.

49. Kandambeth, S., Shinde, D. B., Panda, M. K., Lukose, B., Heine, T., Banerjee, R. (2013). *Angew. Chem. Int. Ed.*, **52**, 13052.

50. Liao, H., Ding, H., Li, B., Ai, X., Wang, C. (2014). *J. Mater. Chem. A*, **2**, 8854.

51. Xu, F., Jin, S., Zhong, H., Wu, D., Yang, X., Chen, X., Wei, H., Fu R., Jiang, D. (2015). *Sci. Rep.*, **5**, 8225–8230.

52. Bai, L., Gao Q., Zhao, Y. (2016). *J. Mater. Chem. A*, **4**, 14106–14110.

53. Kuhn, P., Antonietti M., Thomas, A. (2008). *Angew. Chem. Int. Ed.*, **47**, 3450–3453.

54. Ding, S., Gao, J., Wang, Q., Zhang, Y., Song, W., Su, C., Wang, W. (2011). *J. Am. Chem. Soc.*, **133**, 19816–19822.

55. Uribe-Romo, F. J., Doonan, C. J., Furukawa, H., Oisaki K., Yaghi, O. M. (2011). *J. Am. Chem. Soc.*, **133**, 11478–11481.

56. Kandambeth, S., Mallick, A., Lukose, B., Mane, M. V., Heine, T., Banerjee, R. (2012). *J. Am. Chem. Soc.*, **134**, 19524–19527.

57. Jackson, K. T., Reich T. E., El-Kaderi, H. M. (2012). *Chem. Commun.*, **48**, 8823–8825.

58. Yang, H., Zhang, S., Han, L., Zhang, Z., Xue, Z., Gao, J., Li, Y., Huang, C., Yi, Y., Liu H., Li, Y. (2016). *ACS Appl. Mater. Interfaces*, **8**, 5366–5375.

59. Jlassi, K., Mekki, A., Zayani, M. B., Singh, A., Aswai D. K., Chehimi, M. M. (2014). *RSC Adv.*, **4**, 65213–65222.

60. Fukuoka, T., Tonami, H., Maruichi, N., Uyama, H., Kobayashi, S., Higashimura, H. (2000). *Macromolecules*, **33**, 9152–9155.

61. Zhan, H., Lamare, S., Ng, A., Kenny, T., Guernon, H., Chan, W. K., Djurisic, A. B., Harvey, P. D., Wong, W. Y. (2011). *Macromolecules*, **44**, 5155–5167.

62. Demir-Cakan, R., Morcrette, M., Nouar, F., Davoisne, C., Devic, T., Gonbeau, D., Dominko, R., Serre, C., F′erey G., Tarascon, J. M. (2011). *J. Am. Chem. Soc.*, **133**, 16154–16160.

63. Zheng, J., Tian, J., Wu, D., Gu, M., Xu, W., Wang, C., Gao, F., Engelhard, M. H., Zhang, J., Liu J., Xiao, J. (2014). *Nano Lett.*, **14**, 2345–2352.

64. Zhou, J., Li, R., Fan, X., Chen, Y., Han, R., Li, W., Zheng, J., Wang, B., Li, X. (2014). *Energy Environ. Sci.*, **7**, 2715–2724.

65. Shi, Y., Chou, S., Wang, J., Wexler, D., Li, H., Liu H., Wu, Y. (2012). *J. Mater. Chem.*, **22**, 16465–16470.

66. Lee, J. W., Lim, S. Y., Jeong, H. M., Hwang, T. H., Kang, J. K., Choi, J. W. (2012). *Energy Environ. Sci.*, **5**, 9889–9894.

67. Zhou, G., Wang, D., Li, F., Zhang, L., Li, N., Wu, Z., Wen, L., Lu, G., Cheng, H. (2010). *Chem. Mater.*, **22**, 5306–5313.

68. Yoo, E., Kim, J., Hosono, E., Zhou, H., Kudo, T., Honma, I. (2008). *Nano Lett.*, **8**, 2277–2282.

69. Sen, S., Moses, K., Bhattacharyya, A. J., Rao, C. N. R. (2014). *Chem. Asian J.*, **9**, 100–103.

70. Hwang, H., Kim, H., Cho, J. (2011). *Nano Lett.*, **11**, 4826– 4830.

71. Li, N., Liu, G., Zhen, C., Li, F., Zhang, L., Cheng, H. (2011). *Adv. Funct. Mater.*, **21**, 1717–1722.

72. Xu, X., Rout, C. S., Yang, J., Cao, R., Oh, P., Shin, H. S., Cho, J. (2013). *J. Mater. Chem. A*, **1**, 14548–14554.

73. Zhu, J., Yang, D., Yin, Z., Yan, Q., Zhang, H. (2014). *Small*, **10**, 3480–3498.

Chapter 6

Biomedical Applications of COFs

Distinctive features, like large surface areas, tunable porosity, and π conjugation of covalent organic frameworks (COFs), undergo unique photoelectric properties and will enable COFs to serve as a promising platform for drug delivery, bioimaging, biosensing, and theranostic applications. In this chapter, the resent advanced research on COFs is introduced in the biomedical and pharmaceutical sectors and the challenges and opportunities regarding COFs are discussed in terms of biomedical purposes. Although currently still in their infancy, COFs as an innovative source have paved a new way to meet future challenges in human healthcare and disease theranostic.

6.1 Introduction of Biomedical Application

Recent decades have witnessed a surge of exploration related to biomedical applications of biosensing [1], bioimaging [2, 3], chemotherapy [4, 5], gene therapy [6–9], immunotherapy [10, 11], photodynamic therapy (PDT) [12], photochemical therapy [13, 14], tissue engineering [15, 16], and others [17]. Multifarious materials have been widely developed to achieve these objects, including organic (liposomes [18–21], polymers [22–24], dendrimers [25, 26], etc.), inorganic (metals [27], metallic oxides [28, 29], carbon [30, 31], mesoporous silica [32–34], etc.), and hybrid [35–38] materials.

Covalent Organic Frameworks
Atsushi Nagai

ISBN 978-981-4800-87-7 (Hardcover), 978-1-003-00469-1 (eBook)
www.jennystanford.com

To further improve their therapeutic efficacy, researchers endow these biomaterials with stimuli responsiveness [39, 40] and targeted delivery [41, 42]. Some agents have been approved by the Food and Drug Administration (FDA) of the United States or in clinical trials, such as Doxul. However, there are still more biomaterials limited to clinical trials due to their unsatisfactory efficiency and safety [43]. Organic materials usually suffer from low in vivo stability and loading capacities [44], while inorganic substances often possess undesirable toxicity and poor degradability [45]. Metal-organic frameworks (MOFs), which are hybrid materials, have high surface areas, large pore sizes, and good biodegradability, but their chemical stability and toxicity are unsatisfactory since they are constructed by metal coordination bonds [46]. These issues need to be addressed. Exploitation of new materials is one of the possible methods.

Covalent organic frameworks (COFs) are crystalline porous polymers with periodic skeletons and large surface areas [47]. Similar to MOFs, COFs have abundant regular pores with controllable sizes and shapes and are easy to modify, but they are linked by dynamic covalent bonds as molecular building blocks, which makes them not only more stable than conventional organic materials because of the molecular self-assembly process but also more adaptive than inorganic particles formed by ionic or metallic bonds [48]. In addition, COFs are often composed of light atoms, avoiding the toxicity of metal ions. This kind of composition also features COFs with low density. For example, the density of COF-18, a typical boronate ester COF, is only 0.17 g/cm^{-3} [49].

Recently, COFs were developed for diagnoses and treatment [50, 51] since they exhibit many excellent properties, including stimuli responsiveness, biodegradability, high surface areas, large pore volumes, tunable pore structures, π-conjugated systems, unique photoelectric properties, outstanding modifiability, and so on. The next section introduces the applications of COFs in drug delivery, photochemical therapy, PDT, biosensing, bioimaging, and others.

6.2 COF Properties of Biomedical Applications

COFs possess multiple prominent advantages that allow them to be used in theranostic applications.

- **Dynamic covalent linkages for stimuli responsiveness and biodegradation**

  Instead of traditional covalent bonds, coordinate bonds, and physical interactions, COFs utilize dynamic covalent bonds as linkages, which can preserve their structures in normal conditions and they can be broken by stimulants such as acids [52]. Their characters make COFs stable enough to reach the target tissues and finish their task before they biodegrade. Moreover, they can rapidly release cargos under stimulation, especially for imine and its analogue linkage frameworks. Polymer-covered boronate ester and spiroborate-linked COFs can also be promising candidates for drug delivery, as the reversibility and the dynamic nature of the boronate ester offer not only pH-responsive and glucose-responsive (dual responsiveness) features but also adaptive mechanical behaviors, such as shear-thinning protonation of injection [53–55]. Recently, covalent triazine frameworks (CTFs) were found to possess pH responsiveness due to the protonation and biodegradation of the triazine units [56]. Apparently, the stimuli responsiveness and biodegradation make COFs ideal carriers for therapy.

- **High BET surface area and large pore volume for high capacity**

  COFs possess high surface areas and large pore volumes because of their unique framework skeletons. For instance, the Brunauer–Emmett–Teller (BET) surface area of COF-102 is 3472 m^2/g and of COF-103 is 4210 m^2/g; the pore volume of COF-103 is 1.66 m^3/g and of COF-102 is 1.44 m^3/g [57, 58]. These features are much better than those of most porous materials, such as mesoporous silica and carbon [59, 60]. Furthermore, COFs are constructed by aromatic compounds and can be used to load aromatic monomers through π–π stacking. Thus, COFs possess a high loading capacity for drugs and biological molecules.

- **Tunable pore character for controlled release**

  As described in Chapter 1, the pore sizes and shapes of COFs can be adjusted by changing the building blocks or after modification. Since pore sizes and shapes have an effect on the diffusion of guest monomers, COFs with tunable pore structures can be used to carry different pharmaceutical agents and control their release rates.

- **Wonderful photoelectronic properties for biosensing and bioimaging**
  Mostly, COFs contain π-conjugated systems and laminated structures, which endow them with excellent photoelectronic properties. For example, polycyclic aromatic units (e.g., tetraphenylethenes and pyrenes) can make COFs emit fluorescence and electron donors and acceptors units (e.g., naphthalimides, porphyrins, and phthalocyanines) can endow COFs with proton conductivity. On the basis of these properties, COFs show a great potential in biosensing and bioimaging.
- **Unique modifiability for versatile functionalities**
  Since COFs consist of building blocks with high tailorability, extensive materials can be designed as building blocks and introduced into their skeletons, enabling COFs to perform as desired, such as producing singlet oxygen for PDT [61]. What's more, tailored attributes of COFs pave the way for the study of structure-activity relationship.
- **Outstanding modifiability for versatile functionalities**
  Modification is an effective way to enhance the applicability of COFs. Inspired by other nanoparticles, surface modification can improve their biocompatibility and targeting ability [62, 63]. Since MOFs can regulate drug release profiles by incorporating different substituents on the pore walls [64], COFs may achieve similar results. Even biological probes and drugs can be linked to COFs by a postsynthesis method. With proper ligands, inorganic pharmaceutical agents (e.g., cisplatin) may be connected to COFs. There are also multifarious possible functionalities waiting for further exploration. These properties open the door for the application of COFs in biomedicine.

6.3 Biomedical COF Applications

Due to unique features that endow them with great potential in diagnosis and treatment, COFs have been explored in drug delivery, photochemical therapy, PDT, biosensing, bioimaging, and other biomedical applications. Even though in their early stages, COFs have attracted a lot of researcher attention and a surge of studies have been reported in the past several years.

6.3.1 Drug Delivery

The first example of application of COFs in drug delivery was reported by Yan and coworkers [65], who designed two new 3D polyimide (PI) COFs with different pore sizes, PI-COF-4 (13 Å) and PI-COF-5 (10 Å). Ibuprofen (IBU; molecular size 5Å × 10 Å) was selected as a model drug as it can be entrapped by the pores of PI COFs. Thermogravimetric analysis (TGA) proved that the drug loading efficacy (DLE) was as much as 24% (PI-COF-4) and 20% (PI-COF-2). What's more, the drug release was sustained for more than 6 days. Besides IBU, PI COFs are also suitable to deliver captopril and caffeine, achieving similar results to that of IBU.

Porphyrin-based covalent triazine frameworks (PCTFs) were also used in drug delivery [66]. The authors believed that the acid group decorated in IBU could interact with the triazine rings of COFs, which would have a positive effect on drug loading and controlled release. TGA illustrated that the DLE was about 19% for metal-free PCTFs and 23% for Mn-coordinated PCTFs. For both PTCFs, the total release amount was beyond 90% within 48 h. These results are no better than those of the above-mentioned PI COFs, which may be because PCTFs have irregular morphologies and wider pore size distribution because of the poor crystallinity of triazine-based COFs [67].

Although a number of studies were conducted to investigate in vitro drug release from various COFs, few works are focused on the in vivo biocompatibility and cytotoxicity. Zhao et al. carried out cell experiments with two imine-linked 2D COFs, PI-2 COFs and PI-3 COFs [68]. Both COFs exhibited high drug loading capacities for 5-FU, captopril, and IBU, even reaching 30 wt%. The release rates of 5-FU were similar for both PI-*n* COFs, and most of the drug was released within 3 days (Fig. 6.1c).

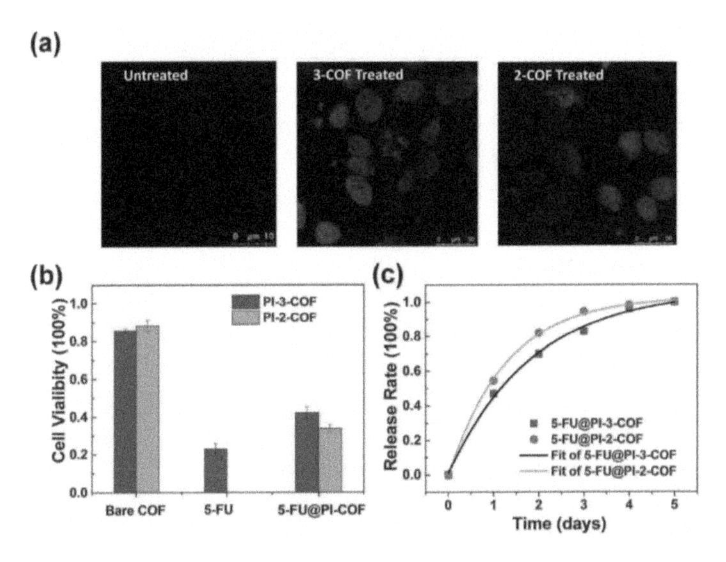

Figure 6.1 (a) Confocal images of cells before and after the treatment of drug-loaded COFs, (b) quantitative MIT analysis showing the cell viability when incubated with bare COFs, 5-FU, and 5-FU-loaded COFs for 24 h. (c) Release profiles of two 5-FU-loaded COFs at 100 mg/mL and corresponding fitting curves. Republished with permission of Royal Society of Chemistry, from Ref. [68], copyright (2016); permission conveyed through Copyright Clearance Center, Inc.

Importantly, these COFs were uniform spherical nanoparticles with a diameter of 50 nm when dispersed in a phosphate-buffered saline (PBS) solution with a bit of dimethyl sulfoxide as an additive. This feature made COFs carriers of a suitable size for cell uptake and in vivo drug delivery. Confocal images proved the effective uptake of drug-loaded nanoparticles by the MCF-7 cells (Fig. 6.1a). Quantitative 3-(4,5-dimethylthiazol-2-yl)-2.5-diphenyltetrazolium bromide (MIT) analysis indicated that both COFs showed good biocompatibility. However, the viability of the cells was decreased to 10% when they were treated with 5-FU-loaded hybrids for 48 h, as shown in Fig. 6.1b. Lotsch and coworkers demonstrated a new imine-based TTI COF to deliver quercetin (Fig. 6.2a,b) [69]. The TTI COF could retain its morphology in water and weak acids for more than two days. Molecular dynamic simulations proved that the quercetin was bound to the pore wall by C–H–π and H bond interactions (Fig. 6.2c,d). For this reason, the TTI COF seemed to be able to carry more drugs. However, trifluoroacetic acid (TFA) analysis failed to calculate the accurate drug loading rate since the quercetin decomposed slowly over a wide temperature range. What's worse, an in vitro drug-release study was also not successful due to the fact that the quercetin was hard to dissolve in PBS and oxidized easily [70]. Cell experiments indicated that the drug-loaded COF successfully killed most of the human breast carcinoma MDA-MB-231 cells while the bare COFs showed no cytotoxicity. Regardless of some inadequacies, this work carefully discussed every detail and shared with us valuable experiences to design this new carrier.

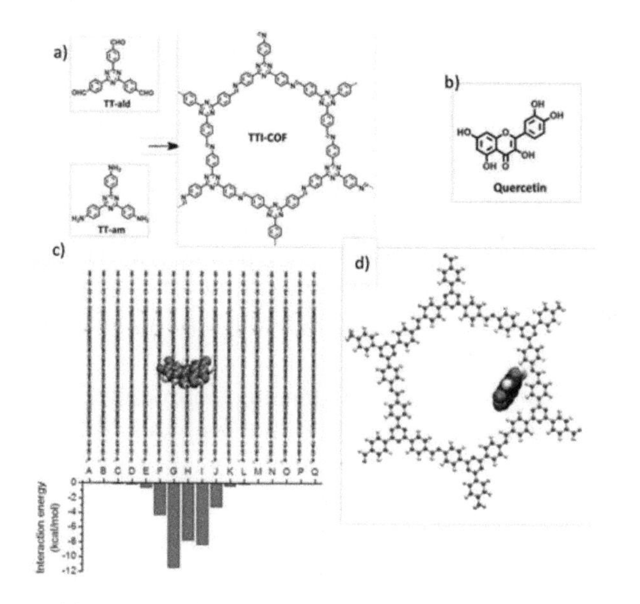

Figure 6.2 (a) Synthesis of a TTI COF from TT-ald and TT-am, (b) structure of quercetin, (c) side view of the modeled COF pore showing interaction energies between the quercetin and the model hexagon layers, and (d) top view of the modeled COF pore showing the interaction of quercetin with the pore wall. Reproduced with permission from Ref. [69]. Copyright (2016), John Wiley and Sons.

Stimuli responsiveness is one of most fascinating properties to achieve real-time monitoring and on-demand drug release [71, 72].

Besides the inherent acid sensitivity of boronate- and imine-based COFs, researchers have endowed COFs with more environment-responsive features to control the drug delivery and release. Huh et al. developed pH-responsive CTF nanoparticles (NCTPs) via a Friedel–Crafts reaction [73]. Doxorubicin (DOX) was loaded in NCTP as a model drug, and the hydrophobic interactions between DOX and NCTP were weakened. Thus, the release rate was faster than that under a PBS solution of pH 7.4. More interestingly, COFs as drug carriers can provide more flexibility and additional option for access to controllable drug release because of their unique photoelectronic properties. For example, Lei and coworkers applied self-condensation of 4,4′-phenylazobenzyl diboronic acid (ABBA) to build single-layered photoresponsive COFs on the surface of highly oriented pyrolytic graphite [74]. Under ultraviolet (UV) irradiation, the frameworks would be destroyed and release the guest, copper phthalocyanine (CuPc). Furthermore, the destroyed COFs could be recovered through annealing. This finding offers a possible on/off switch for drug release.

In 2015, Banerjee and coworkers reported a series of explorations of imine COFs for drug loading and release [75–77]. They prepared hollow spherical COFs via a template-free method [76]. The drug loading capacity was only 0.35 mg/g for DOX, calculated by UV-visual absorbance spectra. The release rate was very slow, and more than 50% of the drug remained after 7 days in a pH 5 phosphate buffer. After that, they designed self-standing porous COF membranes to sieve larger molecules (more than 1 nm) such as rose Bengal, tetracycline, and curcumin [77]. Since the enhanced permeability and retention (EPR) effect is not effective as expected and only about 5% of the administered nanoparticles can reach the target tumor, modifying with target groups is a possible way to improve the use of nanocarriers [42]. Banerjee and coworkers took the first step regarding COFs for targeted drug delivery (Fig. 6.3). They modified covalent organic nanosheets with folic acid (naming it "TpASH-FA") through the postsynthetic modification method, and 5-FU was chosen as the model drug [75]. However, the loading efficiency of these nanosheets (only 12%) was lower than that of many other COFs [78–81]. They exhibited pretty good anticancer activity, and only 14% of MDA-MB-231 cells survived under a dosage of 50 μg/mL.

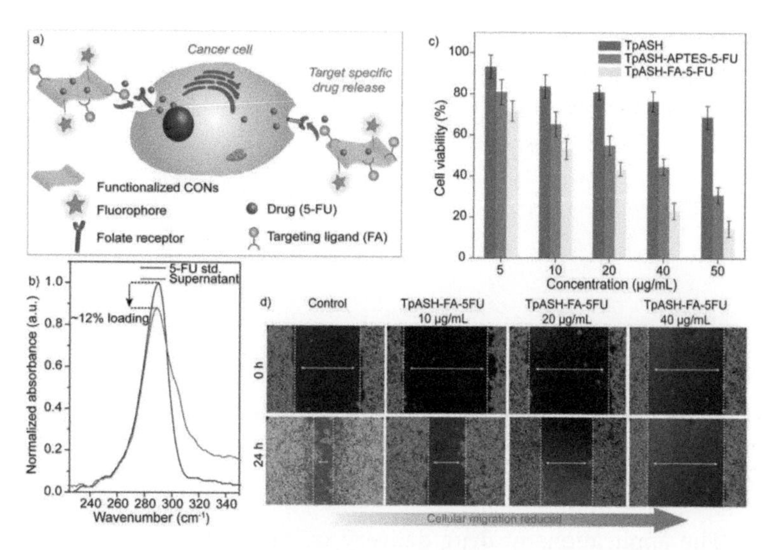

Figure 6.3 (a) Schematic representation of targeted drug delivery by CONs (sheet-like material denotes CONs here), (b) drug loading study of 5-FU by UV-vis spectra, (c) MTT assay on MDA-MB-231 cell lines showing cellular viability, and (d) comparison of cellular migration study between control and TPASH-FA-5-FU-treated sets. Reprinted with permission from Ref. [75]. Copyright (2017) American Chemical Society.

Apart from carrying anticancer drugs, functional COFs can kill tumor cells by themselves. Bhaumik et al. reported an interesting work in which phloroglucinol-contained COFs were utilized as anticancer agents [82]. They applied 2,4,6-triformylphloroglucinol and 4,4′-ethylenediamine to construct nanofiber-like COFs (EDTFP-1) with a diameter of 22–30 nm, as shown in Fig. 6.4a. The phloroglucinol derivative that was created could accelerate reactive oxygen species generation and caused the apoptosis of cancer cells [83]. This cell experiment indicated that EDTFP-1 successfully induced mitochondrial-dependent apoptosis associated with DNA fragmentation, mitochondrial membrane potential loss, phosphatidylserine externalization, and pro- and antiapoptotic protein imbalance, as shown in Fig. 6.4b. Thus, EDTFP-1 showed obvious cytotoxicity against cancer cells, like HCT 116, HepG2, A549, and MIA-Paca2. This finding will open a door for COFs to work as anticancer agents.

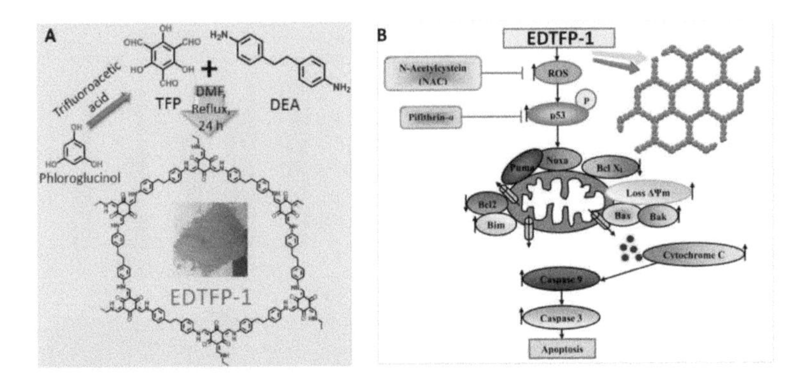

Figure 6.4 Schematic representations of the preparation of EDTFP-1 (A) and its induced apoptotic pathway (B). Reprinted with permission from Ref. [82]. Copyright (2017) American Chemical Society.

The applications of drug delivery using COFs are summarized in Table 6.1. As per the table, most drug-loading COFs are linked by imine (C–N) bonds, which are believed to be stable in water. Regardless, for COF_{ABBA} built by boroxine bonds, neither loading efficacy nor release rate was studied. The pore sizes are mostly designed to fit the model drug, but these carriers have varied instead of uniform nanoparticles, which needs further exploration. In general, the DLE of COFs was rather high and the drugs and the COFs showed good biocompatibility. Several limitations markedly hamper their clinical translation. However, COFs have shown their pharmaceutical potential. The intricate structural characteristics demand careful engineering of the COFs in order to realize the desirable effect. The complexity of the COF preparations is a key issue that needs to be addressed to scale up the production or to ensure batch-to-batch reproducibility. Challenges still exist in terms of delivery of the cargo to the targeted site as well as efficient clearance of the COFs once they have finished their mission in vivo. Although several studies have tested the efficacy of COF formulation and their safety, few studies have been carried out on their long-term accumulation and degradation profiles. However, I believe that continuous improvement in the design and detailed investigations on their in vivo behavior will help to tale COFs as nanocarriers from bench to besides.

Table 6.1 Examples of COFs as drug carriers

Year	COFs	Linkages	Pore sizes (nm)	Morphologies	Model drug	Characters	Ref.
2015	PI-COF-4 PI-COF-5	Imide	1.3 1.0	Rectangular; length hundreds of nanometers	IBU, caffeine, and captopril	DLE[a]: 24 wt %; DRR: 95% for 6 days. DLE[a]: 20 wt %; DRR: 95% for 6 days.	[65]
2015	COF-DhaTab	Imine	3.7	Submicron hollow spheres	DOX	DLE[b]: 0.35 mg/g; DRR: 42% after 7 days at pH 5.	[76]
2016	PI-2 COF PI-3 COF	Imine	1.4 1.1	Spherical nanoparticles; 50 nm	5-FU, IBU, and captopril	DLE[a]: 30 wt %; DRR: 85% for 5 days; good biocompatibility.	[82]
2016	COF_{ABBA}	Boroxine	2.1	Single layer	CuPc	Photoresponsive release; no DLE or DRR measured.	[74]
2016	TTI COF	Imine	2.4	Elongated rods	Quercetin	DLE[b]: Failed to measure; DRR: Failed to measure; good biocompatibility.	[69]

(*Continued*)

Table 6.1 (*Continued*)

Year	COFs	Linkages	Pore sizes (nm)	Morphologies	Model drug	Characters	Ref.
2016	NCTP	Triazine	1.21	Spherical nanoparticles; 50–70 nm	DOX	DLE[a]: 20 wt%; DRR: 60% for 2 days at pH 7.4 and 80% for 2 days at pH 4.8; good biocompatibility.	[79]
2017	EDTFP-1	β-ketoenamine	1.5	Nanofibers; diameter 22–30 nm	TFP	TFP works as a model drug as well as a building block; cancer cells killed by COFs themselves.	[82]
2017	TpASH-FA	β-ketoenamine	1.3	Nanosheets	5-FU	DLE[b]: 12 wt%; DRR: 50% for 75 h at pH 7.4 and 75% for 75 h at pH 5; good biocompatibility.	[75]
2017	PCTF PCTF-Mn	Triazine	0.8–2.7 0.74.2	Irregular nanoparticles; plate shaped	IBU	DLE[a]: 19 wt%; DRR: 90% for 48 h. DLE[a]: 23 wt%; DRR: 95% for 48 h.	[80]

[a] Measured by TFA.
[b] Measured by UV.
DLE; drug loading efficacy; DDR: drug release rate; IBU: ibuprofen; DOX: doxorubicin; 5-FU: 5-fluorouracil; CuPc: copper phthalocyanine; TFP: 2,4,6-triformylphloroglucinol.

6.3.2 Photothermal and Photodynamic Therapy

Besides chemotherapy, COFs can also be utilized for photothermal therapy (PTT) [84] and PDT [85, 86].

Prompted by conjugated microporous polymers [87], Guo et al. attempted to cover imine-linked COFs on the surface of Fe_3O_4 nanoclusters for PTT, as shown in Fig. 6.5a [84]. The resulting Fe_3O_4@COF microspheres increased the system temperature by 25°C (Fig. 6.5b), and the photochemical conversion efficiency (21.5%) was comparable to that of some widely studied photosensitizers, such as Au nanorods [88, 89]. Moreover, the Fe_3O_4 core featured the microspheres with magnetic target characteristic.

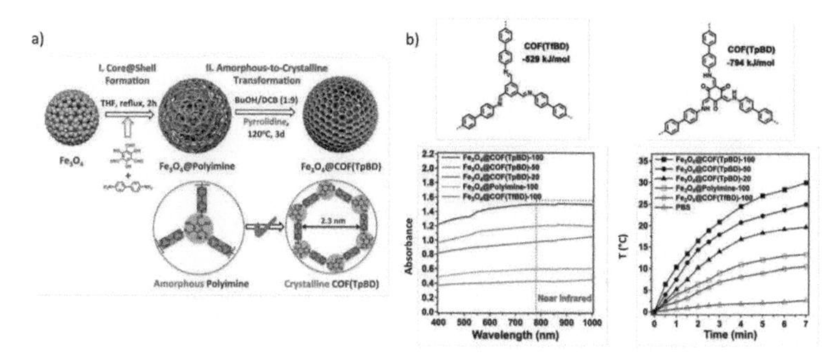

Figure 6.5 Schematic representations of the preparation (a) and photochemical effect (b) of Fe_3O_4@COF. Reproduced with permission from Ref. [84]. Copyright (2016), John Wiley and Sons.

Porphyrin and its derivatives can generate singlet oxygen ($^1O_2$) under photoirradiation and are widely studied as photosensitizers for PDT [90]. Using porphyrin derivatives as building blocks and 3D structures, the construction of photosensitive COFs is a promising method to avoid the aggregation (such as H- or J-aggregates) of porphyrin macrocycles in aqueous media and improve their efficiency in vivo. Wang et al. designed 3D porphyrin-based COFs with excellent photosensitivity, resulting in 3D effects enabling the suppression of intermolecular and intramolecular π–π stacking interaction. Authors could be used to provide reactive singlet oxygen under photoirradiation [86]. The photosensitive properties of COFs could be tuned by the incorporation of metal ions into the center of the porphyrin macrocycle. Xie and coworkers grew imine-linked COFs on the surface of an amine-decorated MOF to prepare porphyrin-based COFs with uniform nanoscale structures, as shown in Fig. 6.6a–c [91]. The amine-modified MOF was a regular octahedron around 165 nm. The hybrid particles (referred to as UNM) were nearly spherical and had an average size of 176 nm, determined by dynamic light scattering. Confocal laser scanning microscopy images showed that the UNM could be endocytosed by HeLa cells. As shown in Fig. 6.6b, with 2′,7′-dichlrodihydrofluorescein diacetate DCFH-DA as indicators, bright-green fluorescence could be observed under light irradiation, which was attributed to the production of singlet oxygen. MTT assay proved that UNM had significant cytotoxicity for both HepG2 and HeLa cells under irradiation but nearly no cytotoxicity can be found without light.

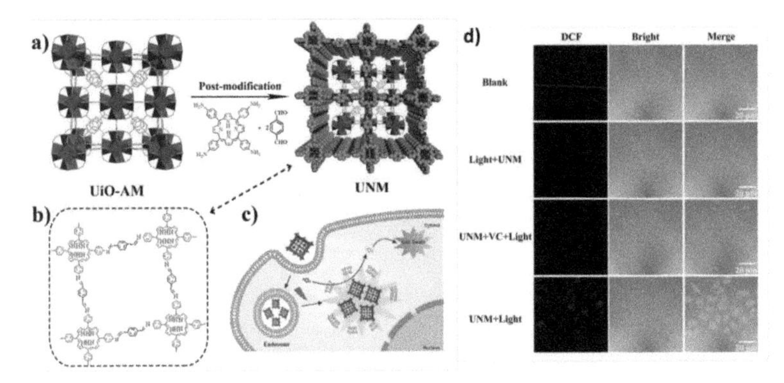

Figure 6.6 (a) Synthesis and preparation of UNM nanoparticles, (b) structure of POP on UNM nanoparticles, (c) the cellular uptake and photodynamic therapy in cells, and (d) the photodynamic effect of MOF@COF. Reprinted with permission from Ref. [91]. Copyright (2017) American Chemical Society.

6.4 Biosensing and Bioimaging

The π-conjugated system and photoelectrical properties of COFs have been widely studied in catalysis [92, 93], proton conduction [94, 95], and energy storage [96, 97]. Lately, these features have been used for biosensing and bioimaging [98–100].

One of the biosensors was designed on the basis of the electrochemical activity of COF films. Fang et al. prepared imine-linked COF films on amino-functionalized silicon wafers, and the presence of amino groups can endow the hybrid films with a positive charge [98]. Negatively charged biomolecules, for example, BSA and probe DNA, would be adsorbed on the surface by electrostatic interactions to strengthen the electrochemical activity of the COF films. Electrochemical impedance spectroscopy was used to detect this change. Thus, the biological signals were converted into electric signals. The other biosensing application is based on the π-conjugated system of COFs [99, 100]. Utilizing the π–π stacking effect, a fluorescent dye was quenched via fluorescence resonance energy transfer [100]. When the target DNA combined with probes, the interaction between the DNA probes and COF films weakened and the fluorescence was recovered. Similarly, a carboxyfluorescein-labeled probe (FAM probe) was adsorbed on the β-ketoenamine-linked COFs by hydrogen bond and π–π stacking interactions [99]. As shown in Fig. 6.7, when the target biomolecules interacted with the FAM probes, the fluorescence was enhanced. This new platform showed highly sensitive and selective DNA and adenosine 5′-triphosphate.

Besides biomolecules, COFs can also be applied to the detection of monomers and metal ions. Jiang and coworkers designed azine-linked pyrene-based COFs in which the azine units extended p delocalization over the 2D skeleton, enabling the framework to emit a yellow light [101]. Furthermore, azine units have lone pairs of electrons and provide docking sites for hydrogen-bonding interaction with guest molecules. Thus, these frameworks can be utilized to detect 2,4,6-trinitrophenol with high sensitivity and selectivity since phenol units can form hydrogen bonds with azine units and nitro groups can quench the fluorescence. Similarly, another 3D pyrene-based COF was used for chemosensing [102]. Furthermore, Yang and coworkers prepared two kinds of polyimide-based COFs that emitted strong fluorescence in solution, and the fluorescence would be quenched by Fe^{3+} [103]. This property can be used for the selective luminescence of Fe^{3+}.

Triazine-based frameworks of NTCTPs showed a strong emission property and can be used for bioimaging due to extended π conjugation in the frameworks [104]. What's more, the aggregation-induced emission (AIE) effect and blue luminescence of conjugated COFs can hopefully be developed for bioimaging [105–107].

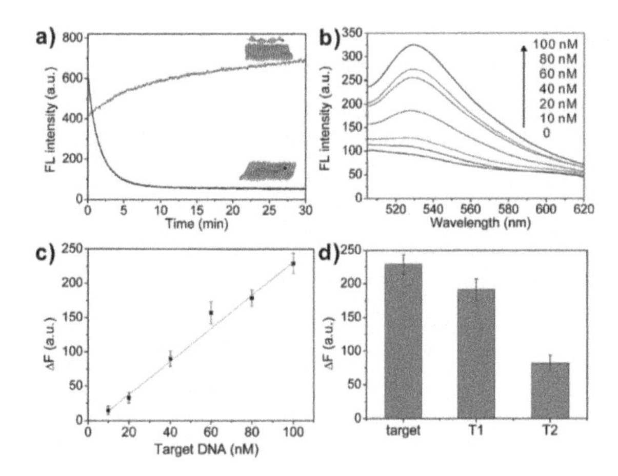

Figure 6.7 (a) Time courses for the fluorescence quenching of the probe (50 nM) upon TpTta addition (10 μg) (black curve) and for the fluorescence recovery of the probe/TpTta upon target DNA addition (400 nM DNA) (red curve). (b) Dependence of the fluorescence of the probe (50 nM) on the concentration of the target DNA (0, 10, 20, 40, 60, 80, and 100 nM). (c) Plot of the recovered fluorescence intensity (DF) against the target DNA concentration. (d) Selectivity of this DNA assay. The fluorescence intensity was measured at 528 nm under excitation at 493 nm. The error bars represent the standard deviation of three replicate measurements. Reproduced of Royal Society of Chemistry, with permission from Ref. [99], copyright (2014); permission conveyed through Copyright Clearance Center, Inc.

6.5 Other Biomedical Applications

With their selective adsorption ability, COFs can be applied in protein immobilization, biomolecular adsorption, and hazardous substance removal. Immobilization of enzymes plays an important role in biomedical industries [108]. Banerjee and coworkers designed hollow spherical COFs (COF-DhaTab) for trypsin immobilization [109]. Although trypsin is a globular protein with a diameter of 3.8 nm, which is a bit larger than the pore size of COF-DhaTab, it could be immobilized due to its soft character. Figure 6.8 shows various methods that were used to demonstrate the immobilization of trypsin by COFs and the loading capacity was as high as 15.5 μmol/g. Cai et al. designed core-shell-structured Fe_3O_4@COFs by way of room temperature synthesis [110]. On the basis of the hydrophobic effect, the composite nanoparticles adsorbed hydrophobic peptides and repelled hydrophilic peptides. With the assistance of the magnetic responsiveness of Fe_3O_4 cores, hydrophobic peptides were separated from the mixture system. Further, these particles were used for the selective enrichment of peptides and exclusion of proteins. This feature has great potential in real-world applications, such as in the analysis of human serum. When amine groups were incorporated onto the pore walls, COFs were used for the adsorption of lactic acid [111]. Other works show that when modified with sulfur derivatives, COFs were used to remove Hg^{2+}, Hg^0, Pd^{2+}, and Cu^{2+} [112, 113].

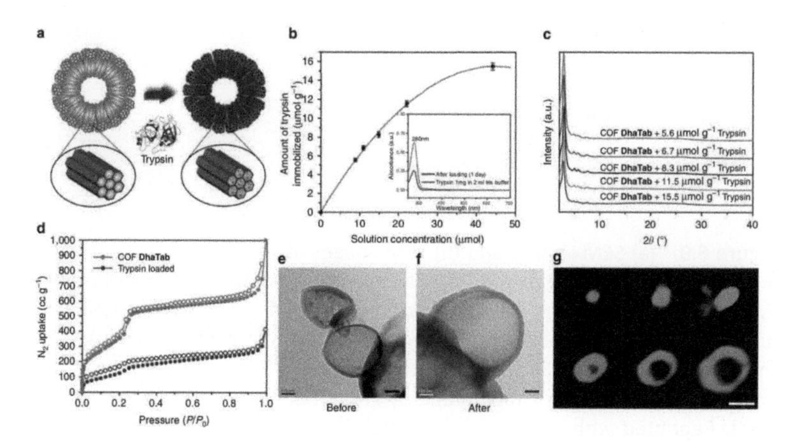

Figure 6.8 (a) Schematic representation of trypsin immobilization, (b) amount of trypsin loading in COF-DhaTab; evidence of trypsin immobilization through (c) PXRD patterns, (d) N_2 adsorption isotherms, (e, f) TEM images, and (g) confocal Z stacks. Reprinted by permission from Springer Nature Customer Service Centre GmbH: Springer Nature, *Nature Communications*, Ref. [109], copyright (2015).

COFs can be used for antimicrobial applications. 1,3,5-triformylphloroglucinol and guanidinium halide were used to build self-exfoliated ionic covalent organic nanosheets (iCONs) [114]. Since guanidinium units can form hydrogen bonds with the oxoanions of phosphate, positively charged iCONs can break the negatively charged phospholipid bilayers of bacteria, as shown in Fig. 6.9. Antibacterial studies testified that these nanosheets show excellent antimicrobial activity against both gram-positive and gram-negative bacteria. Furthermore, they mixed iCONs and polysulfone

(PSF) to fabricate an iCONs@PSF membrane as an antimicrobial coating. Moreover, COFs are expected to deliver bioactive gases, like nitric oxide (NO), as MOFs do [115].

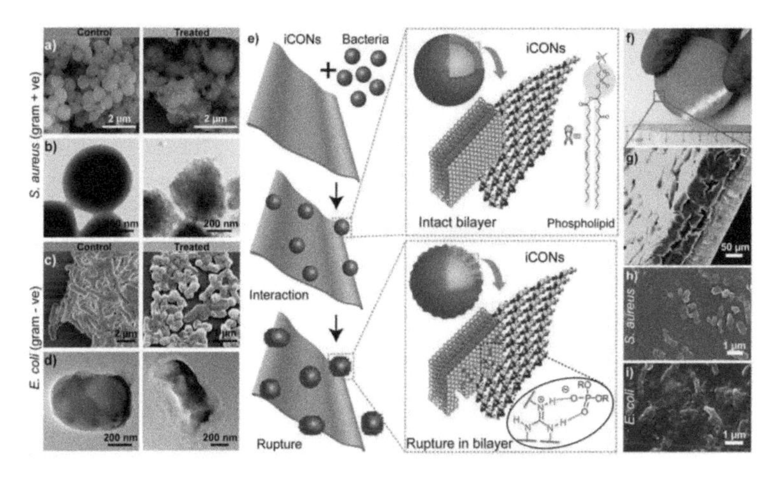

Figure 6.9 (a) SEM images and (b) TEM images of control and $TpTG_{Cl}$-treated *S. aureus*. (c) SEM images and (d) TEM images of control and $TpTG_{Cl}$-treated *E. coli*. (e) Schematic representation for the mode of action between bacteria and iCONs. (f) Digital image of a $TpTG_{Cl}$@PSF mixed matrix membrane. (g) SEM image of the $TpTG_{Cl}$@PSF mixed matrix membrane. Antibacterial property of $TpTG_{Cl}$@PSF mixed matrix membrane by growth of (h) *S. aureus* and (i) *E. coli* on it. Reprinted with permission from Ref. [114]. Copyright (2016) American Chemical Society.

6.6 Conclusions of Biomedical Applications

COFs, a representative material of porous organic polymers, have been widely explored in various fields and play an important role in nanotechnology. In this chapter, some biomedical examples of COFs were introduced to pave the way for their porous frameworks to be used for biomedical purposes. Most importantly, the complementary functional design of skeletons and high flexibility in the control of pore geometry may provide a powerful means of exploring COFs for challenging biomedicine issues. Specifically because of the unique photoelectric properties, COFs are expected to provide an ideal theranostic platform since it is easy to integrate the drug therapy and photoelectric diagnostic function in one COF-based system.

So, it is no wonder that though in its infancy, COFs' application in drug delivery, PTT, PDT, biosensing, bioimaging, and other possible diagnosis has significantly increased and received increasing attention with the development of new COF materials. However, despite their many favorable characteristics and great progress, most of the currently available COFs face many challenges in terms of preparation complexity, regulated morphology, hydrolytic stability, targeted delivery, controlled release, and long-term biocompatibility, which indeed impairs their practical application in biomedical and pharmaceutical fields. As a result, considerable effort is required to design and fabricate COFs with well-defined morphologies and desirable properties. Especially, future work may need to focus on the following in the biomedical field:

- Scaling up the preparation of COFs with a controlled and uniform morphology
- Ensuring hydrolytic and hemodynamic stability
- Ensuring biocompatibility and addressing systemic toxicity

Although there is still a long way to go before COFs can be applied in clinical settings, the rapid development of these materials will help them become a potential platform for biomedical and pharmaceutical purposes, opening up some exciting new avenues to improve human welfare.

References

1. Kim, S.-J., Choi, S.-J., Jang, J.-S., Cho, H.-J., Kim, I.-D. (2017). *Acc. Chem. Res.*, **50**, 1587–1596.
2. Wolfbeis, O. S. (2015). *Chem. Soc. Rev.*, **44**, 4743–4768.
3. Smith, B. R., Gambhir, S. S. (2017). *Chem. Rev.*, **117**, 901–986.
4. Huang, P., Liu, J., Wang, W., Zhang, Y., Zhao, F., Kong, D., Liu, J., Dong, A. (2016). *Acta Biomater.*, **40**, 263–272.
5. Huang, P., Wang, W., Zhou, J., Zhao, F., Zhang, Y., Liu, J., Liu, J., Dong, A., Kong, D., Zhang, J. (2015). *ACS Appl. Mater. Interfaces*, **7**, 6340–6350.
6. Riley, M. K., Vermerris, W. (2017). *Nanomaterials*, **7**, 94.
7. Zhou, J., Wu, Y., Wang, C., Cheng, Q., Han, S., Wang, X., Zhang, J., Deng, L., Zhao, D., Du, L. (2016). *Nano Lett.*, **16**, 6916–6923.
8. Wang, C., Du, L., Zhou, J., Meng, L., Cheng, Q., Wang, C., Wang, X., Zhao,

D., Huang, Y., Zheng, S. (2017). *ACS Appl. Mater. Interfaces*, **9**, 32463–32474.

9. Tatiparti, K., Sau, S., Kashaw, S. K., Iyer, A. K. (2017). *Nanomaterials*, **7**, 77.
10. Li, P., Zhou, J., Huang, P., Zhang, C., Wang, W., Li, C., Kong, D. (2017). *Regen. Biomater.*, **4**, 11–20.
11. Li, P., Song, H., Zhang, H., Yang, P., Zhang, C., Huang, P., Kong, D., Wang, W. (2017). *Nanoscale*, **9**, 13413–13418.
12. Croce, R., Van Amerongen, H. (2014). *Nat. Chem. Biol.*, **10**, 492–501.
13. Cherukula, K., Manickavasagam Lekshmi, K., Uthaman, S., Cho, K., Cho, C. S., Park, I. K. (2016). *Nanomaterials*, **6**, 76.
14. Pattani, V. P., Tunnell, J. W. (2012). *Lasers Surg. Med.*, **44**, 675–684.
15. Amezcua, R., Shirolkar, A., Fraze, C., Stout, D. A. (2016). *Nanomaterials*, **6**, 133.
16. Chieruzzi, M., Pagano, S., Moretti, S., Pinna, R., Milia, E., Torre, L., Eramo, S. (2017). *Nanomaterials*, **6**, 134.
17. Song, G., Cheng, L., Chao, Y., Yang, K., Liu, Z. (2017). *Adv. Mater.*, **29**, 1700996.
18. Deng, H., Song, K., Zhao, X., Li, Y., Wang, F., Zhang, J., Dong, A., Qin, Z. (2017). *ACS Appl. Mater. Interfaces*, **9**, 9315–9326.
19. Lombardo, D., Calandra, P., Barreca, D., Magazu, S., Kiselev, M. A. (2016). *Nanomaterials*, **6**, 125.
20. Campani, V., Salzano, G., Lusa, S., De Rosa, G. (2016). *Nanomaterials*, **6**, 131.
21. Al-Jamal, W. T., Kostarelos, K. (2011). *Acc. Chem. Res.*, **44**, 1094–1104.
22. Wang, Y., Li, P., Truong-Dinh Tran, T., Zhang, J., Kong, L. (2016). *Nanomaterials*, **6**, 26.
23. Amirmahani, N., Mahmoodi, N. O., Mohammadi Galangash, M., Ghavidast, A. (2017). *J. Ind. Eng. Chem.*, **55**, 21–34.
24. Gong, J., Chen, M., Zheng, Y., Wang, S., Wang, Y. (2012). *J. Control. Release*, **159**, 312–323.
25. Kesharwani, P., Lyer, A. K. (2015). *Drug Discov. Today*, **20**, 536–547.
26. Sikwal, D. R., Kalhapure, R. S., Govender, T. (2017). *Eur. J. Pharm. Sci.*, **97**, 113–134.
27. Boisselier, E., Astruc, D. (2009). *Chem. Soc. Rev.*, **38**, 1759–1782.
28. Pankhurst, Q., Thanh, N., Jones, S., Dobson, J. (2009). *J. Phys. D: Appl. Phys.*, **42**, 224001.

29. Ulbrich, K., Hola, K., Subr, V., Bakandritsos, A., Tucek, J., Zboril, R. (2016). *Chem. Rev.*, **116**, 5338–5431.

30. Mendes, R. G., Bachmatiuk, A., Büchner, B., Cuniberti, G., Rümmeli, M. H. (2013). *J. Mater. Chem. B*, **1**, 401–428.

31. Hong, G., Diao, S., Antaris, A. L., Dai, H. (2015). *Chem. Rev.*, **115**, 10816–10906.

32. Martinez-Carmona, M., Colilla, M., Vallet-Regi, M. (2015). *Nanomaterials*, **5**, 1906–1937.

33. He, Q., Shi, J., Chen, F., Zhu, M., Zhang, L. (2010). *Biomaterials*, **31**, 3335–3346.

34. Tang, F., Li, L., Chen, D. (2012). *Adv. Mater.*, **24**, 1504–1534.

35. Wu, M.-X., Yang, Y.-W. (2017). *Adv. Mater.*, **29**, 1606134.

36. Lismont, M., Dreesen, L., Wuttke, S. (2017). *Adv. Funct. Mater.*, **27**, 1606314.

37. He, C., Liu, D., Lin, W. (2015). *Chem. Rev.*, **115**, 11079–11108.

38. Zhao, F., Yao, D., Guo, R., Deng, L., Dong, A., Zhang, J. (2015). *Nanomaterials*, **5**, 2054–2130.

39. Liu, X., Yang, Y., Urban, M. W. (2017). *Macromol. Rapid Commun.*, **38**, 1700030.

40. Li, F., Lu, J., Kong, X., Hyeon, T., Ling, D. (2017). *Adv. Mater.*, **29**, 1605897.

41. Bar-Zeev, M., Livney, Y. D., Assaraf, Y. G. (2017). *Drug Resist. Updates*, **31**, 15–30.

42. Toy, R., Bauer, L., Hoimes, C., Ghaghada, K. B., Karathanasis, E. (2014). *Adv. Drug Deliv. Rev.*, **76**, 79–97.

43. Bobo, D., Robinson, K. J., Islam, J., Thurecht, K. J., Corrie, S. R. (2016). *Pharm. Res.*, **33**, 2373–2387.

44. Blanco, E., Shen, H., Ferrari, M. (2015). *Nat. Biotechnol.*, **33**, 941–951.

45. Deng, C., Jiang, Y., Cheng, R., Meng, F., Zhong, Z. (2012). *Nano Today*, **7**, 467–480.

46. Yang, P., Gai, S., Lin, J. (2012). *Chem. Soc. Rev.*, **41**, 3679–3698.

47. Huxford, R. C., Della Rocca, J., Lin, W. (2010). *Curr. Opin. Chem. Biol.*, **14**, 262–268.

48. Huang, N., Wang, P., Jiang, D. (2016). *Nat. Rev. Mater.*, **1**, 16068.

49. Yaghi, O. M. (2016). *J. Am. Chem. Soc.*, **138**, 15507–15509.

50. El-Kaderi, H. M., Hunt, J. R., Mendoza-Cortés, J. L., Côté, A. P., Taylor, R. E., O'Keeffe, M., Yaghi, O. M. (2007). *Science*, **316**, 268–272.

51. Fang, Q., Wang, J., Gu, S., Kaspar, R. B., Zhuang, Z., Zheng, J., Guo, H., Qiu, S., Yan, Y. (2015). *J. Am. Chem. Soc.*, **137**, 8352–8355.

52. Mitra, S., Sasmal, H. S., Kundu, T., Kandambeth, S., Illath, K., Diaz Diaz, D., Banerjee, R. (2017). *J. Am. Chem. Soc.*, **139**, 4513–4520.

53. Mura, S., Nicolas, J., Couvreur, P. (2013). *Nat. Mater.*, **12**, 991–1003.

54. Chen, J., Su, Q., Guo, R., Zhang, J., Dong, A., Lin, C., Zhang, J. (2017). *Macromol. Chem. Phys.*, **218**, 1700166.

55. Guo, R., Su, Q., Zhang, J., Dong, A., Lin, C., Zhang, J. (2017). *Biomacromolecules*, **18**, 1356–1364.

56. Zhao, F., Wu, D., Yao, D., Guo, R., Wang, W., Dong, A., Kong, D., Zhang, J. (2017). *Acta Biomater.*, **64**, 334–345.

57. El-Kaderi, H. M., Hunt, J. R., Mendoza-Cortés, J. L., Côté, A. P., Taylor, R. E., O'Keeffe, M., Yaghi, O. M. (2007). *Science*, ***316***, 268–272.

58. Côté, A. P., El Kaderi, H. M., Furukawa, H., Hunt, J. R., Yaghi, O. M. (2007). *J. Am. Chem. Soc.*, **129**, 12914–12915.

59. Hong, G., Diao, S., Antaris, A. L., Dai, H. (2015). *Chem. Rev.*, **115**, 10816–10906.

60. Martinez-Carmona, M., Colilla, M., Vallet-Regi, M. (2015). *Nanomaterials*, **5**, 1906–1937.

61. Lin, G., Ding, H., Chen, R., Peng, Z., Wang, B.: Wang, C. (2017). *J. Am. Chem. Soc.*, **139**, 8705–8709.

62. Bar-Zeev, M., Livney, Y. D., Assaraf, Y. G. (2017). *Drug Resist. Updates*, **31**, 15–30.

63. Greenwald, R. B., Choe, Y. H., McGuire, J., Conover, C. D. (2003). *Adv. Drug Deliv. Rev.*, **55**, 217–250.

64. Dong, Z., Sun, Y., Chu, J., Zhang, X., Deng, H. (2017). *J. Am. Chem. Soc.*, **139**, 14209–14216.

65. Fang, Q., Wang, J., Gu, S., Kaspar, R. B., Zhuang, Z., Zheng, J., Guo, H., Qiu, S., Yan, Y. (2015). *J. Am. Chem. Soc.*, **137**, 8352–8355.

66. Luo, Y., Liu, J., Liu, Y., Lyu, Y. (2017). *J. Polym. Sci. Part A: Polym. Chem.*, **55**, 2594–2600.

67. Kuhn, P., Forget, A., Su, D., Thomas, A. (2008). *J. Am. Chem. Soc.*, **130**, 13333–13337.

68. Bai, L., Phua, S. Z., Lim, W. Q., Jana, A., Luo, Z., Tham, H. P., Zhao, L., Gao, Q., Zhao, Y. (2016). *Chem. Commun.*, **52**, 4128–4131.

69. Vyas, V. S., Vishwakarma, M., Moudrakovski, I., Haase, F., Savasci, G., Ochsenfeld, C., Spatz, J. P., Lotsch, B. V. (2016). *Adv. Mater.*, **28**, 8749–8754.

70. Wybranowski, T., Kruszewski, S. (2014). *Acta Phys. Pol. A*, **125**, A57–A60.

71. Mura, S., Nicolas, J., Couvreur, P. (2013). *Nat. Mater.*, **12**, 991–1003.

72. Lu, Y., Aimetti, A. A., Langer, R., Gu, Z. (2016). *Nat. Rev. Mater.*, **2**, 16075.

73. Rengaraj, A., Puthiaraj, P., Haldorai, Y., Heo, N. S., Hwang, S. K., Han, Y. K., Kwon, S., Ahn, W. S., Huh, Y. S. (2016). *ACS Appl. Mater. Interfaces*, **8**, 8947–8955.

74. Liu, C., Zhang, W., Zeng, Q., Lei, S. (2016). *Chemistry (Easton)*, **22**, 6768–6773.

75. Mitra, S., Sasmal, H. S., Kundu, T., Kandambeth, S., Illath, K., Diaz Diaz, D., Banerjee, R. (2017). *J. Am. Chem. Soc.*, **139**, 4513–4520.

76. Kandambeth, S., Venkatesh, V., Shinde, D. B., Kumari, S., Halder, A., Verma, S., Banerjee, R. (2015). *Nat. Commun.*, **6**, 6786.

77. Kandambeth, S., Biswal, B. P., Chaudhari, H. D., Rout, K. C., Kunjattu, H. S., Mitra, S., Karak, S., Das, A., Mukherjee, R., Kharul, U. K., Banerjee, R. (2017). *Adv. Mater.*, **29**, 1603945.

78. Mitra, S., Sasmal, H. S., Kundu, T., Kandambeth, S., Illath, K., Diaz Diaz, D., Banerjee, R. (2017). *J. Am. Chem. Soc.*, **139**, 4513–4520.

79. Rengaraj, A., P uthiaraj, P., Haldorai, Y., Heo, N. S., Hwang, S. K., Han, Y. K., Kwon, S., Ahn, W. S., Huh, Y. S. (2016). *ACS Appl. Mater. Interfaces*, **8**, 8947–8955.

80. Luo, Y., Liu, J., Liu, Y., Lyu, Y. (2017). *J. Polym. Sci. Part A: Polym. Chem.*, **55**, 2594–2600.

81. Bai, L., Phua, S. Z., Lim, W. Q., Jana, A., Luo, Z., Tham, H. P., Zhao, L., Gao, Q. (2016). *Chem. Commun.*, **52**, 4128–4131.

82. Bhanja, P., Mishra, S., Manna, K., Mallick, A., Das Saha, K., Bhaumik, A. (2017). *ACS Appl. Mater. Interfaces*, **9**, 31411–31423.

83. Zhang, Y., Luo, M., Zu, Y., Fu, Y., Gu, C., Wang, W., Yao, L., Efferth, T. (2012). *Chem. Biol. Interact.*, **199**, 129–136.

84. Tan, J., Namuangruk, S., Kong, W., Kungwan, N., Guo, J., Wang, C. (2016). *Angew. Chem. Int. Ed. Engl.*, **55**, 13979–13984.

85. Zheng, X., Wang, L., Pei, Q., He, S., Liu, S., Xie, Z. (2017). *Chem. Mater.*, **29**, 2374–2381.

86. Lin, G., Ding, H., Chen, R., Peng, Z., Wang, B., Wang, C. (2017). *J. Am. Chem. Soc.*, **139**, 8705–8709.

87. Tan, J., Wan, J., Guo, J., Wang, C. (2015). *Chem. Commun.*, **51**, 17394–17397.

88. Pattani, V. P., Tunnell, J. W. (2012). *Lasers Surg. Med.*, **44**, 675–684.

89. Tan, J., Wan, J., Guo, J., Wang, C. (2015). *Chem. Commun.*, **51**, 17394–17397.

90. Croce, R., Van Amerongen, H. (2014). *Nat. Chem. Biol.*, **10**, 492–501.

91. Zheng, X., Wang, L., Pei, Q., He, S., Liu, S., Xie, Z. (2017). *Chem. Mater.*, **29**, 2374–2381.

92. Kamiya, K., Kamai, R., Hashimoto, K., Nakanishi, S. (2014). *Nat. Commun.*, **5**, 5040.

93. Lin, S., Diercks, C. S., Zhang, Y.-B., Kornienko, N., Nichols, E. M., Zhao, Y., Paris, A. R., Kim, D., Yang, P., Yaghi, O. M. (2015). *Science*, **349**, 1208–1213.

94. Xu, H., Tao, S., Jiang, D. (2016). *Nat. Mater.*, **15**, 722–726.

95. Chandra, S., Kundu, T., Kandambeth, S., Babarao, R., Marathe, Y., Kunjir, S. M., Banerjee, R. (2014). *J. Am. Chem. Soc.*, **136**, 6570–6573.

96. Chandra, S., RoyChowdhury, D., Addicoat, M., Heine, T., Paul, A., Banerjee, R. (2017). *Chem. Mater.*, **29**, 2074–2080.

97. Deng, W., Li, Y., Zheng, S., Liu, X., Li, P., Sun, L., Yang, R., Wang, S., Wu, Z., Bao, X. (2017). *Angew. Chem. Int. Ed.*, **139**, 8194–8199.

98. Wang, P., Kang, M., Sun, S., Liu, Q., Zhang, Z., Fang, S. (2014). *Chin. J. Chem.*, **32**, 838–843.

99. Li, W., Yang, C. X., Yan, X. P. (2017). *Chem. Commun.*, **53**, 11469–11471.

100. Peng, Y., Huang, Y., Zhu, Y., Chen, B., Wang, L., Lai, Z., Zhang, Z., Zhao, M., Tan, C., Yang, N. (2017). *J. Am. Chem. Soc.*, **139**, 8698–8704.

101. Dalapati, S., Jin, S., Gao, J., Xu, Y., Nagai, A., Jiang, D. (2013). *J. Am. Chem. Soc.*, **135**, 17310–17313.

102. Guo, J., Xu, Y., Jin, S., Chen, L., Kaji, T., Honsho, Y., Addicoat, M. A., Kim, J., Saeki, A., Ihee, H. (2013). *Nat. Commun.*, **4**, 2736

103. Wang, T., Xue, R., Chen, H., Shi, P., Lei, X., Wei, Y., Guo, H., Yang, W. (2017). *New J. Chem.*, **41**, 14272–14278.

104. Rengaraj, A., Puthiaraj, P., Haldorai, Y., Heo, N. S., Hwang, S. K., Han, Y. K., Kwon, S., Ahn, W. S., Huh, Y. S. (2016). *ACS Appl. Mater. Interfaces*, **8**, 8947–8955.

105. Dalapati, S., Jin, E., Addicoat, M., Heine, T., Jiang, D. (2016). *J. Am. Chem. Soc.*, **138**, 5797–5800.

106. Wan, S., Guo, J., Kim, J., Ihee, H., Jiang, D. (2008). *Angew. Chem. Int. Ed. Engl.*, **47**, 8826–8830.

107. Baldwin, L. A., Crowe, J. W., Shannon, M. D., J aroniec, C. P., McGrier, P. L. (2015). *Chem. Mater.*, **27**, 6169–6172.

108. Lei, C., Shin, Y., Liu, J., Ackerman, E. J. (2002). *J. Am. Chem. Soc.*, **124**, 11242–11243.

109. Kandambeth, S., Venkatesh, V., Shinde, D. B., Kumari, S., Halder, A., Verma, S., Banerjee, R. (2015). *Nat. Commun.*, **6**, 6786.

110. Lin, G., Gao, C., Zheng, Q., Lei, Z., Geng, H., Lin, Z., Yang, H., Cai, Z. (2017). *Chem. Commun.*, **53**, 3649–3652.

111. Lohse, M. S., Stassin, T., Naudin, G., Wuttke, S., Ameloot, R., DeVos, D., Medina, D. D., Bein, T. (2016). *Chem. Mater.*, **28**, 626–631.

112. Sun, Q., Aguila, B., Perman, J., Earl, L. D., Abney, C. W., Cheng, Y., Wei, H., Nguyen, N., Wojtas, L., Ma, S. (2017). *J. Am. Chem. Soc.*, **139**, 2786–2793.

113. Huang, N., Zhai, L., Xu, H., Jiang, D. (2017). *J. Am. Chem. Soc.*, **139**, 2428–2434.

114. Mitra, S., Kandambeth, S., Biswal, B. P., Khayum, M. A., Choudhury, C. K., Mehta, M., Kaur, G., Banerjee, S., Prabhune, A., Verma, S. (2016). *J. Am. Chem. Soc.*, **138**, 2823–2828.

115. Xiao, B., Wheatley, P. S., Zhao, X., Fletcher, A. J., Fox, S., Rossi, A. G., Megson, I. L., Bordiga, S., Regli, L., Thomas, K. M. (2007). *J. Am. Chem. Soc.*, **129**, 1203–1209.

Index